STRONG FORCE

WOMEN'S ADVENTURES IN SCIENCE

STRONG FORCE
the story of physicist
SHIRLEY ANN JACKSON

by Diane O'Connell

Franklin Watts
A Division of Scholastic Inc.
New York • Toronto • London • Auckland • Sydney
Mexico City • New Delhi • Hong Kong
Danbury, Connecticut

Joseph Henry Press
Washington, D.C.

AUTHOR'S ACKNOWLEDGMENTS

I am deeply grateful to you, Dr. Shirley Ann Jackson, for taking time out of your impossible schedule to meet with me many times throughout the writing of this project. Your intelligence, persistence in spite of obstacles, passion for your work, compassion for others, and "big picture thinking" have been a true inspiration. Thank you also for teaching me about physics and making the incomprehensible understandable. Thanks also to the dedicated staff at RPI for all their assistance: Patrice DeCoster, Melissa Hogan, Tracey Leibach, and Lucy Norman. Thank you to all those who kindly agreed to be interviewed: Barbara Avery, Mary K. Gaillard, Paul Gray, Anthony Johnson, Gloria Joseph, and Cynthia R. McIntyre. Thanks also to John Quackenbush and Laura Farra for educating me in the finer points of physics. Finally, I thank my husband, Larry, for your cheerleading, great judgment, creative guidance, strong shoulders, and unwavering belief in me. This one's for you. — DO'C

Cover photo: Physicist Shirley Ann Jackson is shown giving a lecture at Rensselaer Polytechnic Institute in New York.

Cover design: Michele de la Menardiere

Library of Congress Cataloging-in-Publication Data

O'Connell, Diane, 1956-
    Strong force : the story of physicist Shirley Ann Jackson / Diane O'Connell.
        p. cm. — (Women's adventures in science)
    Includes bibliographical references and index.
    ISBN 0-531-16784-4 (lib. bdg.) ISBN 0-309-09553-0 (trade pbk.) ISBN 0-531-16959-6 (classroom pbk.)
1. Jackson, Shirley Ann, 1946- 2. Physicists—United States—Biography—Juvenile literature.
3. Women physicists—United States—Biography—Juvenile literature. I. Title. II. Series.

    QC16.J33O36 2005
    530'.092—dc22

                                                                                    2005000827

Any opinions, findings, conclusions, or recommendations expressed in this volume are those of the author and do not necessarily reflect the views of the National Academy of Sciences or its affiliated institutions.

Printed in the United States of America.
1 2 3 4 5 6 7 8 9 10 R 14 13 12 11 10 09 08 07 06 05

About the Series

The stories in the *Women's Adventures in Science* series are about real women and the scientific careers they pursue so passionately. Some of these women knew at a very young age that they wanted to become scientists. Others realized it much later. Some of the scientists described in this series had to overcome major personal or societal obstacles on the way to establishing their careers. Others followed a simpler and more congenial path. Despite their very different backgrounds and life stories, these remarkable women all share one important belief: the work they do is important and it can make the world a better place.

Unlike many other biography series, *Women's Adventures in Science* chronicles the lives of contemporary, working scientists. Each of the women profiled in the series participated in her book's creation by sharing important details about her life, providing personal photographs to help illustrate the story, making family, friends, and colleagues available for interviews, and explaining her scientific specialty in ways that will inform and engage young readers.

This series would not have been possible without the generous assistance of Sara Lee Schupf and the National Academy of Sciences, an individual and an organization united in the belief that the pursuit of science is crucial to our understanding of how the world works and in the recognition that women must play a central role in all areas of science. They hope that *Women's Adventures in Science* will entertain and enlighten readers with stories of intellectually curious girls who became determined and innovative scientists dedicated to the quest for new knowledge. They also hope the stories will inspire young people with talent and energy to consider similar pursuits. The challenges of a scientific career are great but the rewards can be even greater.

Contents

Exploring the Unseen

Some explorers lift off into space. Others dive deep beneath the sea or trek through mountains or rain forests. Shirley Ann Jackson explores the universe too, but on a smaller scale and from a different perspective. She is a physicist who investigates the world on a submicroscopic level. She studies the tiniest elements of the universe, the particles that make up all matter.

Why is it so exciting to learn about things you can't even see? Split one of those tiny particles and you could cause catastrophic destruction. Or you could harness its powerful force to improve the way we work and play and live our lives.

Because of Shirley's scientific expertise and her willingness to take on new challenges, her career has moved in exciting directions. In industry, she worked on the cutting edge of technology, discovering ways for materials to be more useful in our everyday lives. When President Bill Clinton asked her to head the Nuclear Regulatory Commission in 1995, she made sweeping changes to ensure the public's safety, in our own country and around the world. As president of Rensselaer Polytechnic Institute, she is creating unique opportunities for the next generation of scientists.

How did Shirley Ann Jackson achieve so much? She began with a curious mind and a passion for uncovering the secrets that lay hidden in the world around us.

To Shirley,
the world was
full of mysteries

and living creatures
provided the clues
that could help solve them.

BEE SECRETS

Shirley Ann Jackson stood in the hot summer sun, patiently staring at a large rosebush in her family's garden. It was still early morning, but already the heat was beginning to build. Shirley could feel a sneeze making its way through her nose. At 10 years old, Shirley was allergic to many things, including the pollen from the flowers she admired so much.

But this day, Shirley wasn't just admiring the garden of her family's home in Washington, D.C. She had important work to do. She was capturing bees.

As a large bumblebee headed for the rosebush, Shirley tensed in anticipation. The bee hovered over the bush, flitting here and there until it found just the right spot. It touched down and nestled itself into the center of a wide-open bloom. Shirley watched its fuzzy body vibrate as it slurped up the flower's nectar.

Carefully—very carefully—she reached down with one hand and closed the petals around the bee. She held her breath slightly, then gently plucked the bloom with her other hand. She could feel the bee's wings beating frantically inside the petals as it tried to escape. But that didn't worry Shirley. She had been capturing bees this way since she was eight years old. By now, she knew exactly what to do.

As a young girl, Shirley *(opposite)* was fascinated by the behavior of bees. The bees pictured above are making honey in a honeycomb.

Honeybees and plants share a symbiotic relationship, which means that they benefit each other. Bees consume flower nectar, and flowers reproduce because they are pollinated by bees traveling from one flower to another.

She dropped the bloom containing the bee into an empty mayonnaise jar. Quickly—before the bee even knew where it was—she screwed on the top, which already had tiny holes punched in it so the bee could breathe.

Shirley carried the jar with her new specimen to the wooden porch in the back of the house. She scooted under the porch where it was dark and cool, a perfect place for her bee collection. When her eyes adjusted to the dim light, Shirley picked out a spot on a ledge where she would put her newest addition. It would sit between two other jars. One held three yellow jackets and a wasp, and the other held a wasp and a bumblebee. The summer was only half over, but she had already collected dozens of bees in jars that her mother had cleaned out for her.

~ Bee Behavior

Some people thought it odd that a young girl would want to collect bees. After all, bees sting! And most of the other kids were afraid of them. Amazingly, Shirley had never been stung. But she wasn't just collecting bees the way some people collect dolls or marbles. She had a purpose: to learn about bee behavior and in that way unlock one of the secrets of nature.

To Shirley, the world was full of mysteries, and living creatures provided the clues that could help solve them. She chose bees to study because they were always buzzing about and she would never run out of specimens. Plus, they were easy to keep in captivity.

Shirley had several questions about bees. For example, she wanted to know how they might behave if they were fed certain things. To find out, she had a different way of capturing them. She waited until the bee was at the edge of a flower petal. Then, holding the jar on one side and the top on the other, she closed the bee into the jar without the flower. She added different foods, such as sugar, for the bees to eat. Then she observed their behavior after they ate the different foods.

At first, she kept the bees with their own kind. Then she wondered how they would behave with other species. She decided to mix them all up: bumblebees with yellow jackets, wasps with bumblebees, yellow jackets with wasps. The bumblebees seemed to be the most aggressive at first, but eventually all the species of bees got along pretty well with each other.

Shirley also wondered how the bees would act if they spent more or less time in the dark. In their spot under the eaves the bees were in the dark a lot of the time. Shirley would bring them out during the day to see if their behavior was different when they were in the light.

Like any good scientist, Shirley kept a detailed log of her observations. And when she analyzed her data she discovered some interesting things. For example, Shirley noticed that under normal circumstances, bees have a circadian-type rhythm—or a pattern of behavior that is repeated every 24 hours. Shirley found that she could change this rhythm by changing how long she kept the bees out of the light. If the bees stayed in the dark under the porch until the middle of the day, they tended to behave as though it were the middle of the night.

The flowers and shrubs around Shirley's home provided an endless source of specimens for her bee research.

~ Lessons from the Bees

Surprisingly, the most important thing the bees taught Shirley was not so much a scientific lesson as a lesson about life itself. Shirley realized that no living thing likes to be in captivity. When the bees were first caught, they would bang against the side of the jar, trying to get out. Over time, they got more and more passive. That's when Shirley knew it was time to let them go. But by then the bees had become so used to their new environment that sometimes they didn't leave right away, even after Shirley opened the jar. It was as though they had given up. So Shirley would leave the jars open for as long as it took for the bees to fly away.

As Shirley got older, she saw the lesson of the bees repeat itself in the human world around her. Shirley realized that, just like the bees, people can easily become conditioned to having their space and their possibilities limited. And just like the bees in captivity, these people may simply stop trying.

Beatrice Cosby Jackson taught her children to read before they went to kindergarten. Pictured here are three of the four Jackson children *(top left to right),* Gloria, George, and Shirley.

Shirley's parents already knew this reality all too well. George H. Jackson, a postal supervisor, and Beatrice Cosby Jackson, a social worker, were African Americans raising a family in the 1950s. They saw the limits placed on black children every day. At that time, segregation laws in the United States meant that blacks were not allowed to use the same facilities as whites. In addition to attending separate schools, African Americans drank from different water fountains, used separate restrooms, sat in the backs of buses, and were often barred from eating in the same restaurants as whites.

Though the Jacksons lived only a few blocks from the Barnard School, a public elementary school, they were forced to send their children to Parkview Elementary, an all-black school a couple of miles away. It was more than an inconvenience. The city did not provide school bus service, so families had to figure out on their own how their children would get to and from school each day.

The fathers on Shirley's block banded together to work out a car-pooling system. One father would take the kids to school in the morning and another would pick them up in the afternoon. Each time, they drove right past the "whites only" school.

Segregation was a reality, but Shirley's parents believed that their four children, Barbara, Shirley, Gloria, and George, should still strive to achieve their potential. Education, they knew, was essential for success. "Aim for the stars, so that you can reach the treetops, and at least you'll get off the ground," their father would urge.

Both parents encouraged any educational activities that interested their children. Beatrice had a lifelong love of literature and read to them every night. One of Shirley's favorite books was a biography of Benjamin Banneker. Born in 1731, the African-American son of a former slave was a self-taught clockmaker, astronomer, and mathematician. When Shirley's hometown of Washington, D.C., was being built, Banneker helped survey the land, plan the streets, and select building sites. Shirley drew inspiration from this man who refused to let the prejudices he suffered hinder his intellectual development.

Benjamin Bannaker's
PENNSYLVANIA, DELAWARE, MARYLAND, AND VIRGINIA
ALMANAC,
FOR THE
YEAR of our LORD 1795;
Being the Third after Leap-Year.

—PRINTED FOR—
And Sold by JOHN FISHER, Stationer.
BALTIMORE.

Segregation laws prohibited blacks from using the same facilities as whites. Restriction signs were everywhere—at water fountains, restrooms, restaurants, and train and bus station waiting rooms.

One of Shirley's heroes was Benjamin Banneker (1731–1806). Among other accomplishments, he published six annual farmer's almanacs. They provided information on tides, medicine and medical treatment, and eclipses that he calculated himself.

~ And They're Off!

While Shirley's mother instilled in her a love of literature, her father shared his talent for mathematics and science. Though he had only a high school education, he had a great thirst for knowledge and a natural ability for math and mechanical things. He had put these gifts to work during World War II. During the Allied landing at Normandy in France, the vehicles that transported the soldiers from water to land lost their rudder mechanisms. This made them vulnerable to enemy fire. Shirley's father fashioned a new steering mechanism for these vehicles from scrap metal, which ensured the safety of the soldiers. For this accomplishment he received a Bronze Star and a special citation.

Now a father, he amazed his children with the mathematical calculations he performed in his head. And after tackling tasks from rebuilding car engines to finishing the family basement, he seemed to have every tool imaginable in his workshop. He encouraged his children to learn to use them.

George Jackson also enjoyed showing his children how to use scientific principles to add some fun to their everyday lives. This included teaching Shirley and her younger sister, Gloria, how to build soapbox go-carts or "hot rods." These wooden racing cars were built from spare parts. They rolled downhill just from the power of a big push.

To build their hot rod, Shirley and Gloria started with wooden planks for the body. Then they went scavenging through the neighborhood for various other parts, such as wheels, pipes, bolts, and pedals. The tricky part was the steering. Shirley tried all kinds of steering mechanisms, but ultimately settled on bicycle handles.

For his special efforts in World War II, Shirley's father, George H. Jackson *(right)*, received a Bronze Star and a special citation from General Benjamin O. Davis, Sr. *(left)*, the first African American general in U.S. history.

Shirley was the second oldest in a family of four siblings: Barbara (in giraffe skirt), Shirley, Gloria, and George.

The finished result looked like a cross between a sled and a wagon.

Shirley, a born leader, organized go-cart races with some of the neighborhood kids. The neighborhood was built on hills, so the drivers could use gravity to propel the carts. They would line up their hot rods in the alleyways that ran between the houses and get ready to race.

Shirley soon figured out that the trick to winning a race was to build a cart that would naturally go faster than the others. This meant that the builder had to be clever in choosing axles and wheels, as well as the bearings around the wheels. Attention also had to be paid to the most important element of the design: the sleekness and shape of the cart's body.

Shirley often spent time reading on the porch of her house in Washington, D.C.

Shirley found that if she made the front of the body narrower than the back, the air would flow around it more freely. This aerodynamic shape reduced air resistance and made the cart go faster. She did not know it at the time, but in designing her go-cart, Shirley was making the same kinds of decisions that car manufacturers make every day. These design decisions were based on physics, the science of matter and energy.

Shirley enjoyed winning the races. What she liked best, though, was figuring out what kinds of materials and design increased speed and efficiency. She also liked to consider where to position the cart before the race in order to get the maximum speed. By doing all this, Shirley was actually learning about applied physics. (*See box.*) She was also learning valuable lessons in leadership.

> Even as a young girl, Shirley had the ability to recover from hurt and to not let obstacles stand in her way.

One particular race taught Shirley another kind of lesson. After Shirley won the race, one of the neighborhood kids became angry. In a jealous rage, he jumped up and down on the Jacksons' cart and broke it in two. Gloria was devastated. Shirley was upset, too, and not just because all their hard work was ruined. She hated to see her family or friends hurt or treated unfairly.

Bodies at Rest, Bodies in Motion

Like everything else on Earth, Shirley's go-carts obeyed the basic principles of physics called Newton's Laws of Motion.

Newton's First Law of Motion includes the idea that an object at rest will remain at rest unless an outside force acts upon it. In the case of the go-cart, Shirley knew her cart would sit at the starting line unless a force—such as gravity, someone giving the cart a push, or both—set it in motion.

Newton's Second Law of Motion describes the relationship between mass and acceleration. This law states that the more mass something has, the more force is needed to change its motion. It also says that the stronger the force, the greater the acceleration.

So what did this mean to Shirley? She knew she couldn't make her go-cart too big or it would be hard to push off the starting line. And she also knew that the starting push would be critical. The bigger the push, the faster the go-cart could move down the hill.

Newton's Third Law of Motion states that for every action there is an equal and opposite reaction. This law would have come into play if Shirley's cart had accidentally veered off course and hit another cart!

On Sundays, Shirley liked to dress up and attend church with her family.

So Shirley considered what she should do. Although she didn't like to back down from a fight, Shirley was smart enough to know that the best revenge was not to get even, but to succeed.

Determined, she went back to her father and asked him to help rebuild the cart. The girls returned another day with the rebuilt cart and went on to race it again and again.

Even as a young girl, Shirley had the ability to recover from hurt and to not let obstacles stand in her way. She also had the power to persevere. This strong inner force and her dignified manner would carry Shirley through other challenges and obstacles in her life.

Besides being
smart, *Shirley was*

a natural **leader.**

AIMING FOR THE STARS

2

Shirley Ann Jackson was born on August 5, 1946, part of the post-World War II "baby boom." During the 1950s, when she was growing up, two historic events had an enormous impact on her life and her future.

The first happened when Shirley was in the third grade. A battle over segregation had been raging for years in the United States. Many people believed that racial segregation of places like schools and restrooms was okay as long as the black facilities and the white ones were equal. In truth, "coloreds only" schools were almost always inferior to the ones for "whites only." On May 17, 1954, the Supreme Court of the United States affirmed that reality. It handed down a decision in one of the most famous cases of the era: *Brown v. the Board of Education of Topeka, Kansas*. The Court held that separate is not equal and ordered schools to integrate their student populations. For Shirley, this meant that she could now attend the Barnard School around the corner, where she would have a chance at a good education.

The second event came a few years later. In the 1950s, the United States and the Soviet Union, rival superpowers, each began stockpiling arsenals of nuclear weapons. Schools all over the United States held air raid drills, where students would file into fortified "air raid shelters" in case of nuclear attack. Then,

In 1964 Shirley *(opposite)* graduated from high school at the top of her class. Barbara Jackson *(above)* and baby sister, Shirley, sit at the Reflecting Pool, a popular attraction near the Washington Monument in Washington, D.C.

In 1954 the U.S. Supreme Court ruled that segregation was unconstitutional. This mother and daughter celebrate the decision on the steps of the Supreme Court.

on October 4, 1957, the Soviet Union launched into space the first unmanned satellite, *Sputnik I.*

When the Soviets beat the Americans into space, it startled the United States into action. The country launched its own space initiative, and the space race was on. That meant that schools all over the country now put a new focus on science and math. Teachers and school administrators looked to identify students who had particular ability in those areas. It was a turning point, not only in our country's history, but for Shirley Jackson as well.

~ Brainiac

At the newly integrated Barnard School, Shirley blossomed intellectually. Unfortunately, her teachers didn't always know what to think of her. During reading circle time in the first grade, students were supposed to read selected passages and then answer questions.

Shirley always zipped through the reading and finished long before her classmates. Whenever the teacher asked a question, Shirley blurted out the answer without waiting to be called on. Her behavior earned her the nickname "Brainiac" by her classmates.

To keep Shirley from disrupting the class, her teacher made her sit in a corner and read by herself. But that didn't stifle her. As the teacher asked the class questions, Shirley just kept shouting out the answers from the corner. By this time, the teacher had had enough. She made a phone call to Shirley's mother.

"Your daughter is disruptive," the teacher told Mrs. Jackson.

"Maybe you aren't giving her enough to do," Shirley's mother replied.

Beatrice Jackson knew her daughter wasn't a troublemaker. She was probably just bored. So she gave Shirley's teacher some advice: "Don't try to force her to read what everybody else is reading or read at the pace everybody else reads at. Give her more to read—more books, more advanced things. Think of questions you want her to answer and she can write them down." Shirley's teacher followed that advice. Shirley not only settled down, she thrived.

In the sixth grade, all the students in the Barnard School were tested and placed in tracks according to their abilities. There was a basic track, business track, college prep, and honors track. Shirley was placed in the honors track, which was the most accelerated.

By the sixth grade, Shirley's academic abilities were recognized and she was placed in an accelerated program.

~ A Clean Sweep

Besides being smart, Shirley was a natural leader. At school she belonged to a group called the Barnard Improvement Council. The group took care of the school and kept it looking nice.

Shirley inspired the kids who lived nearby to care about their neighborhood. Her Clean-Up Campaign produced positive results for everyone involved.

In the summers Shirley extended that concept to her own neighborhood, calling it the Clean-Up Campaign. She marshaled the neighborhood kids to join her in her mission. She handed out brooms that had nails on the end of the handles. The kids speared trash with the nail end and swept up leaves and other debris with the broom end. Shirley organized everyone according to where they lived on the block and made them responsible for sweeping up the area around their houses. Everyone pitched in, putting larger trash into bags and sweeping junk into piles. The piles were then swept up by the street-sweeper trucks that came by once a week.

The Clean-Up Campaign was a big hit with the adults on the block, and the kids had fun working together. The best part for everyone was watching the noisy trucks gobble up the piles of trash, leaving behind a neat-looking street.

~ Role Models

In school, one of Shirley's favorite subjects was Latin. She studied this ancient Roman language from seventh through twelfth grade. By the time she got to her senior year in high school, she had finished all her formal classes in Latin.

Her twelfth-grade Latin teacher, Norma Davenport, was special to Shirley. She understood Shirley's need to immerse herself in learning and designed special assignments just for her. Shirley would read stories and articles in Latin and then translate them into English, and vice versa. Shirley spent a lot of time with Mrs. Davenport. She also spent a lot of time on her porch reading Latin and translating. Shirley loved the challenge of the language. It was never a chore or a struggle. For her, it was as much fun as reading a good novel.

But it was her math class that would ultimately determine Shirley's future. The honors math classes were very small, with only 7 to 10 students. This gave the students and the teacher a chance to interact much more than in the larger classes. Students worked on problems in class, shared their solutions with each other, argued over different approaches, and discussed the uses of mathematics. Here, Shirley flourished.

Shirley had the same math teacher from the tenth through the twelfth grade, Marie Smith. Mrs. Smith's passion for math shone through in the way she talked about it. Math was more than numbers; it was a way of seeing the world and making sense of it. Mrs. Smith recognized Shirley's natural ability for math and found extra problems for Shirley to solve on her own, which Shirley happily did.

Shirley enjoyed everything about math: the logic, the way equations balanced on both sides of the equal sign, being able to ask a question and get a definitive answer, and solving word problems. She was especially intrigued to find patterns and hidden connections. Like her teacher, Shirley saw mathematical equations as a kind of special language. A mathematician can take a big idea that might take many words or paragraphs to explain. By using numbers, letters, and symbols instead, a mathematician can condense it down to its basic truth. An equation communicates exactly the right information to understand a problem—no more, no less.

Mrs. Davenport, Mrs. Smith, and Shirley's economics teacher, Mrs. Eleanor Blackburn, became her role models. These African American women were smart, well-educated, worldly, and dedicated. They inspired students like Shirley to seize whatever opportunities they were granted in life.

Several teachers at the Barnard School served as positive role models for Shirley. Her Latin teacher, Norma Davenport *(top)*, mathematics teacher Marie Smith *(middle)*, and economics teacher, Eleanor Blackburn *(bottom)*, were all inspirational.

~ Busy as a Bee

Shirley also explored her love of both the sciences and language arts through a rich variety of activities beyond the classroom. She participated in school science fairs and attended educational lectures on weekends. She also competed in school and citywide spelling bees and oratorical contests. In fact, Shirley made the semifinal round in the citywide spelling bee, and won an oratorical contest sponsored by her church, the Vermont Avenue Baptist Church. She was also a delegate to the Student United Nations.

> The bright lights, TV cameras, and studio audience didn't faze Shirley. She was used to being in competitions.

Shirley had another chance to demonstrate her knowledge when she appeared on a television program called *It's Academic*. This quiz show brought teams from Washington, Maryland, and Virginia together to compete in a game-show format. Competition was fierce, and students put in many hours of practice and study to prepare for the show. Shirley's team consisted of herself and two classmates. Her team members specialized in English and history questions, while Shirley took care of the science and math.

Shirley participated in many activities beyond the classroom. She even made her own clothes, such as the outfit she is wearing in this photo.

The show was broadcast live from a TV studio. The bright lights, TV cameras, and studio audience didn't faze Shirley. She was used to being in competitions. Instead, she focused on the questions and answered as fast as she could. It turned out that those days in reading circle, calling out rapid-fire answers to the teacher's questions, had come in handy after all.

~ The TOPs

Though Shirley was a phenomenal student, she was more than just a "brainiac." She had many friends at school and belonged to a club of girls called the TOPs, which stood for Teens of Personality. On Fridays each girl wore a uniform of a green pleated skirt, cream-colored blazer, and a green beanie to school. They wanted to be dressed alike for their weekly meeting, held after school at one of the members' homes.

After the meeting, they invited boys over for a "social" and a dance, which had been pre-arranged with their parents. If the weather was cool, they met inside. If the weather was warm, they gathered in the backyard. Everyone had a great time, listening and dancing to their favorite Motown artists. Shirley was an especially good dancer. Her favorite was the jitterbug, a popular, jazzy, couples dance.

Shirley also sang in the choir at the Vermont Avenue Baptist Church, where she had many friends. The church gave her a strong spiritual foundation that she would call upon often throughout her life.

Shirley *(center)* receives one of several scholarships that helped her attend MIT.

~ Applying to MIT

There was never any question that Shirley would go to college. Both her parents had instilled in her the value of education, stressing its importance since early childhood. Midway through her junior year in high school, Shirley began to grapple with the idea of choosing a college.

One of her high school activities was working in the principal's office. During lunch hours and after school Shirley helped out by filing and doing other clerical tasks for the principal, the guidance counselor,

WINNERS—W. F. Lawton, Grand Master of the Most Worshipful Prince Hall Grand Lodge of the District, left, presents high school seniors Michael Williams (Dunbar), Kermit Frazier (Ballou), Shirley Jackson (Roosevelt), Sheila Neustadt (Coolidge), Carrie Palmer (McKinley) and Catherine Peaks (Cardozo), with scholarships on behalf of the Lodge. School Supt. Carl Hansen is at right. The scholarships, which are given annually totaled $10,000.

and the assistant principals. Though it didn't pay, it was an honor to be asked to do this job. And Shirley liked the work.

One day, the assistant principal for boys, Robert Boyd, and the guidance counselor, S. O. Brown, called Shirley into the principal's office. Mr. Brown told her that she would probably be the valedictorian, or top student, of her graduating class.

"You have a lot of ability," Mr. Boyd added. "Have you ever thought about going to MIT?"

Shirley had never heard of the school.

Mr. Brown explained, "It's a science and engineering-based school that takes kids who are exceptional, particularly those who do very well in math." Both men encouraged Shirley to apply to the school and offered to help her fill out the application.

In addition to Shirley's Martin-Marietta scholarship, support for college came from closer sources, like her family and fellow church members, who contributed to a scholarship fund for her.

After that meeting, Shirley ran home, eager to tell her parents the exciting news. In a rush of words, she told her mother and father, "There's this special school that the assistant principal thinks I should apply to and I never heard of it— it's called MIT—have you heard of it?"

Her father had heard of the Massachusetts Institute of Technology, often called MIT for short. "I've read a lot about that school," he told Shirley. "That's where you should go."

Everyone rallied around Shirley. Mrs. Smith, Shirley's math teacher, also urged her to apply. For the assistant principal, the guidance counselor, and the math teacher, Shirley embodied the promising future for African Americans.

Shirley applied and was accepted to MIT. Because of her high level of achievement, she received a four-year scholarship covering tuition and some other expenses from the Martin-Marietta Corporation (now Lockheed Martin), a defense contractor.

MISS SHIRLEY JACKSON
(Scholarship winner)

To brilliant Shirley Ann Jackson of Washington, D.C., last week, went the coveted 1964 Martin Marietta Corporation Foundation College scholarship. The winner will have $1,100 applied to tuition expenses next fall and both will attend the Massachusetts Institute of Technology. Miss Jackson will major in physics and Hendricks is enrolled in electrical engineering. She took the academic-honors curriculum at Roosevelt High School and was ranked first in a class of 320. She was president of the National Honor Society, an office assistant at her school, was secretary of the Wider Horizons Club, co-editor of the yearbook and president of the Latin Club. As a hobby, she has been interested in collecting live bees.

18

Each year the corporation awarded a scholarship to one student from Maryland, one from Virginia, and one from the District of Columbia who were interested in studying math and science. Shirley also received scholarships from the Prince Hall Masons (an African American Masonic lodge) and the Vermont Avenue Baptist Church.

~ "It's My Life"

Despite all the excitement, Shirley's mother wasn't entirely sold on MIT. When Shirley had applied, her mother didn't know much about the school. After she read up on it, she realized that it was one of the best colleges in the country for science and math. But she still had misgivings. For one thing, it was in Boston, far from family and friends in Washington, D.C.

Even more troubling was that only a handful of black students had ever attended the university—and none of them were women. Shirley's mother worried that her daughter would be alone and isolated. She also worried about her daughter's safety. At the time, Bostonians were being terrorized by the Boston Strangler, a serial killer who targeted women.

After Shirley's acceptance, Beatrice Jackson sat down with her husband George and their daughter. She expressed her worries, but Shirley stood her ground. "It's one of the best schools in the country," she argued. "Plus I have these scholarships, which will help pay for it." Shirley also explained that she wanted to get away from Washington so she could experience more of the country. "I need the chance to see if I can make my own way," she said.

But Beatrice was still reluctant to let her daughter go.

"I really want to do this," Shirley insisted, even though she could see her mother was still resisting. "It's my life," she added.

Amid preparations for college, Shirley still managed to enjoy herself. Here, she and her date head off to the high school prom.

That simple phrase touched Beatrice. She realized her daughter did need to make her own way, so she finally gave in. "You're right," she said. "It is your life."

~ Higher Levels

Earning top honors out of more than 300 seniors, valedictorian Shirley Jackson receives her high school diploma during graduation ceremonies at Theodore Roosevelt High School.

On graduation day in June 1964, Shirley stood onstage in the auditorium of Theodore Roosevelt High School. As valedictorian, she was expected to give a speech. She was used to speaking before a crowd, so she wasn't nervous—just excited. Her speech was a longer version of the one with which she won the oratorical contest. Because it had particular meaning to her, she had decided to use it as her valedictory speech.

Before she began speaking, she looked into the audience for the two most important people in her life: her parents. Her speech, which she called "Living on Higher Levels: Opportunities Unlimited," was as much a tribute to them as it was an inspiration to the students she was addressing.

Shirley spoke of living a life focused on doing the right thing, working as hard as one could, and aiming to reach the stars. Doing all these things, she stressed, would create opportunity. As it turned out, the speech would be a road map for how she would live her life.

~ On Her Own

In September Shirley's parents drove her to MIT to help her settle in. It was a long, eight-hour drive from Washington, D.C., to Cambridge, Massachusetts. They all stayed the night at the Hotel Elliot, then went over to the campus the next morning.

This was Shirley's first visit to the campus. As she toured the grounds, it seemed huge. There were many classic buildings and

sculptures by famous artists. The McLaurin Building, with its large dome and columns, was especially inspiring. She couldn't wait to attend classes and lectures inside it. Her parents escorted her to McCormick Hall, the new dormitory for women. Shirley was lucky. If she had gone to MIT only a year earlier, she would have had to live off campus. Until very recently, women had been a rarity at MIT. Now there were nearly 100, enough to fill a dorm of their own.

Shirley thought the new dorm was beautiful. She had a single room, which meant she didn't have to share with a roommate. Her parents helped her get her room set up, then they said goodbye.

"You're leaving?" Shirley asked. She wasn't at all prepared for the final moment of goodbye.

Her parents reassured her that she would do just fine. They wished her well, told her to work hard and take care of herself, and to stay in touch.

After they left, Shirley felt alone for the first time in her life. She sat down on her bed and cried. She didn't know what lay ahead for her at MIT. All she knew was that she was one of only two African American women in a freshman class of nearly a thousand students. She would soon find out just how alone she was.

At MIT, Shirley was impressed by the classic architectural style of the McLaurin Building *(top)* and the women's dormitory, McCormick Hall *(inset)*.

She had thought that
by helping her **classmates,**
she might eventually get them

to see beyond her *race*
and like her **for who she was.**

SHUT OUT

3

When Shirley began her freshman year at MIT in 1964, African Americans were just beginning to make progress in achieving equality. Due in large part to the civil rights movement led by Reverend Dr. Martin Luther King, Jr., African Americans were now being admitted into colleges like MIT that had previously been closed to them. The doors of opportunity were creaking open, but the doors of acceptance to the white community were often shut tight.

Shirley quickly found out that being admitted to MIT and being accepted by her classmates were not the same thing. In class she found herself surrounded by empty seats. Her classmates would sit anywhere, it seemed—except next to her. No one talked to her. For the first time in her life, she felt like she had no friends.

Shirley sought out and befriended the only other African American woman in her freshman class, Jennifer Rudd. But this friendship often made Shirley feel even more like an outsider. Though they looked nothing alike, the other students often mistook one girl for the other. A typical exchange went something like this:

"Hi, Jenny!" a student would say to Shirley.

"My name is not Jenny," Shirley would reply.

Shirley *(opposite center)* plays around with classmates at MIT, performing the "Speak no evil, see no evil, hear no evil" pantomime. The Reverend Dr. Martin Luther King, Jr., *(above)* leads thousands of civil rights advocates and ordinary citizens during the March on Washington in 1963.

"Oh, you're the other."

"Yes, I'm the other one."

Or, if Shirley sat down at a table in the cafeteria where other students were eating, someone would ask her, "Where's Jenny?" And some of them would get up and leave when she sat down, despite the fact that they weren't finished eating. They did this even if all the other tables were full. Shirley couldn't understand why they were acting this way, and it hurt a lot. She had never experienced this kind of cruelty before, not even when the Barnard School had just been integrated.

Although MIT had a small number of women students in 1964, Shirley was one of only two African American women in the freshman class.

~ Study Groups

Shirley immersed herself in her coursework, losing herself in the comfort and familiarity of math and science. She would work on her problem sets in physics alone in her room, her door shut against the noise in the hall.

One night Shirley needed to use the restroom across the hall. She opened her door and stepped out of her room, right into a hallway filled with girls. Papers and textbooks littered the floor. Some of the girls intensely quizzed each other, while others worked feverishly on physics problems. *Study groups!* thought Shirley. Until that moment, she had no idea they existed at MIT.

Shirley remembered Mrs. Smith's high school math class and how much she enjoyed working with the other students on problem sets, debating approaches, and sharing solutions. It was part of what she liked so much about mathematics. So after she finished in the restroom, Shirley went back to her dorm room, gathered up her homework, and stepped back into the corridor. She approached the group of girls.

"May I join you?" she asked.

One of the girls looked up at Shirley. "Go away," she said.

"But I've done half of the problems and you can see my solutions, and I think I know how to do the rest of them," Shirley replied.

Another girl eyed Shirley coldly. "Didn't you hear what she said? She said go away."

Shirley was stung. She turned around and went back into her room. She dropped her papers onto her desk and burst into tears. After she had cried herself out, a determined Shirley said to herself, *Well, I've got to finish the problem set.* And that's exactly what she did.

Shirley later found out that students got together all the time to work on problem sets and study for exams. But they never invited Shirley to any of these sessions. Instead, Shirley had to figure out everything on her own.

Gradually, however, Shirley's classmates began to see that she could be an asset to them. When the tests and papers were passed

> Another girl eyed Shirley coldly. "Didn't you hear what she said? She said go away."

At MIT in 1964, white students sitting with black students at meals was a rare sight.

back and they saw Shirley's high grades, some of the other students began to talk to her. Still, their conversations never went beyond physics.

Shirley resented that her interactions with her classmates were so limited. She was working just as hard as they were and was just as smart, yet they kept her at arm's length. She had thought that by helping her classmates, she might eventually get them to see beyond her race and like her for who she was.

~ *Hanging In*

MIT may not have been very welcoming, and Shirley certainly missed her family back home, but she never considered giving up. She was excelling in her classes. She remembered how important her education was to her parents. Her mother worked and her father had two jobs so that they could help her financially.

She also knew how important it was to the men, women, and children of the Vermont Avenue Baptist Church back home.

Shirley found the dorm rooms, like this double room in McCormick Hall, beautiful and spacious. During her freshman year, her single room served as a retreat where she could study and listen to her favorite music.

They had all contributed to a scholarship for her to attend MIT, and they wanted very much for her to succeed. To them, Shirley was a trailblazer who might help lower barriers for other African Americans in the future. Shirley knew she couldn't let them down.

Shirley also had a strong belief in God. She often read the Bible. The passages about forgiveness had particular meaning for her in the face of the hurtful words and actions of the other students. She also found solace in music. Sometimes she would retreat to her room and listen to her records. The music of her favorite recording artists—the Temptations, the Four Tops, Aretha Franklin, and Marvin Gaye—had a soothing

26

effect on her. As she listened to their uplifting songs, her hurt seemed to fade away.

Most of all, Shirley refused to give up because if she did, it meant that those who wanted her to quit would win. Her determined nature carried her through even the toughest times.

~ Reaching Out

Shirley had been raised with the idea that her own success wasn't enough. It was vital for her to help others, too. So she decided to volunteer at Boston City Hospital in the pediatric ward. There, she found children of all races. Some suffered from grave illnesses, such as leukemia. Others had orthopedic problems that made it difficult for them to walk.

> In many ways, she realized, she was the lucky one. She had her health. She had ability. And most of all, she had opportunity.

Though her heart went out to all these children, one little boy particularly touched her. He had blond hair and a sunny disposition. Unfortunately, to all who saw him for the first time, he was a horror. He had no face. He had sunken sockets where his eyes should have been, a deformity for a nose, and just an opening for the mouth. As severe as his physical deformities were, however, they could eventually be fixed with plastic surgery. Still, no one ever seemed to visit this boy. He seemed to be all alone in the world.

At the beginning of every volunteer shift, Shirley went straight to the little boy's room and held him. Sometimes she went early so she could hold him for a while before she took up her other duties. She sat with him, offering him comfort. She, too, received comfort in return. The boy helped her realize that everyone has a cross to bear of some kind, and that suffering comes to all regardless of race or gender. In many ways, she realized, she was the lucky one. She had her health. She had ability. And most of all, she had opportunity.

~ Sorority Sisters

As much as Shirley's volunteer work meant to her, she still craved a social life. Clearly, MIT wasn't going to offer her one. So she joined a sorority, a national service and social club for women college students. Shirley joined Delta Sigma Theta, one of the oldest African American sororities in the United States.

Through the New England chapter, called Iota, Shirley got to know other African American women from Northeastern University, Boston University, Radcliffe, Brown, Amherst, Wellesley, and Yale, among other schools. Shirley found comfort and strength in getting to know these women who struggled with the same issues she was struggling with.

The sorority also showed Shirley how an enterprising spirit can help you overcome obstacles. Universities may have been admitting more African Americans, but their policies toward them were still hostile. For example, no school would let the African American sorority use their facilities to meet. That meant that the sorority had to find places to hold their functions and pay for every expense—unlike the campus-based sororities.

The Iota chapter found two places that would rent facilities to them. One was a YMCA in Roxbury, a black community in Boston where the sorority had a tutoring program. The other was a club for African American professional businesswomen called the Negro Professional and Business Women's Club.

> Shirley found comfort and strength in getting to know these women who struggled with the same issues she was struggling with.

The sorority invited young men from the surrounding schools to their social events. They charged admission for these events to cover the costs of renting the space and buying refreshments themselves. They also had to pay off-duty policemen to be there because the city required it for all gatherings of a certain size. Had they been allowed to meet on campus, this would not have been an issue. Though the situation was not fair, it taught Shirley and her sorority sisters to be resourceful.

~ Tutoring Math

The sorority was about more than social events. Delta Sigma Theta's mission also included social outreach. So after about a year and a half of volunteering at Boston City Hospital, Shirley turned her attention to tutoring high school students in algebra, trigonometry, and geometry at the Roxbury YMCA. Tutoring gave Shirley the opportunity to work on math problems. Plus, she knew that if she could explain the math problems to others, it meant she understood them herself. But what really got Shirley excited was seeing the breakthrough her students experienced when they finally understood a solution that they had previously thought impossible.

Tutoring, volunteering at the hospital, participating in her sorority, listening to her favorite music, and reading the Bible all helped bolster Shirley's inner confidence. These things gave her the motivation to keep going despite her difficulties. *No one ever said it would be easy,* she often reminded herself.

Still, in spite of the hurtful times, Shirley recognized that there were good things about MIT. She had world-class professors, challenging classes, and the opportunity to learn about a subject she found endlessly fascinating.

Then she took
a class *that opened*
her eyes

to a whole new way
of looking *at the world.*

4

PANIC AT MIT

When Shirley began her freshman year at MIT, she hadn't yet decided on her major. Math was her first love, so she knew that whatever she ended up doing would involve mathematics. But she also was thinking about studying electrical engineering. Then she took a class that opened her eyes to a whole new way of looking at the world.

The class was called "Physics: A New Introductory Course," and it was required of all freshmen. It didn't take long for the students to figure out that the first letters of each word in the course's name spelled PANIC. This nickname did not inspire confidence, but the professor, Anthony French, quickly won over even the most reluctant students. Shirley was impressed by his passion for physics and his energetic style of teaching. She was also intrigued by his British accent!

The class was a lecture and held about 500 students. How do you get the attention of 500 students, some of whom are only taking your class because they have to? You use lots of visual aids and humor. That's just what Professor French did.

In good weather, some classes were held in the Great Court (*opposite*) on MIT's campus. The photo above shows a model of the structure of atoms in a solid.

~ A Roll of the Dice

One lecture in particular was a favorite among many of Professor French's students. It was a discussion of randomness and probability. A central rule of modern physics (quantum mechanics) is that you cannot predict how atoms will act in any particular experiment. (Atoms are the tiny particles, or bits, that make up all matter.) (*See box, page 35.*) You can only predict the likelihood that something will happen based on the results of many, many experiments.

After enlisting a volunteer, Professor French held up a wastepaper basket. He shook it and turned it over. Out rolled a pair of huge wooden dice. The professor looked at the number of dots that came up on each die: two and five.

"Move seven steps to the right," he instructed the volunteer. Professor French repeated the dice roll. This time they came up three and one. "Move four steps to the left." Another time: four and two. "Move six steps to the left," he instructed. He continued the demonstration, directing the volunteer based on the roll of the dice.

By moving his volunteer around, Professor French was trying to give the class a lesson in random behavior. Like any "game of chance," the numbers that come up on the dice can never be predicted with certainty. This randomness is a basic law of quantum mechanics, the study of the tiniest parts of matter. You can never say for sure where any particle of matter will be at a particular moment. You can only make an educated guess about how it will most likely behave.

Professor Anthony French's enthusiastic attitude was infectious and helped students appreciate the wonder of physics.

For Shirley, the world was like a good mystery book with fascinating secrets. Her early experiments with bees allowed her to uncover some of the interesting secrets of life science. Now physics

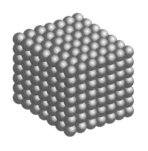
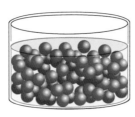
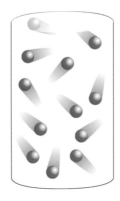

Atoms are constantly in motion, and the way that they move determines what objects look like and how they behave.

Solid
In a solid, atoms vibrate against each other but are held in tightly packed rows. A solid keeps its shape unless an outside force changes it.

Liquid
In a liquid, atoms are close together but they slide over each other. This movement allows liquids to take the shape of the container they are in.

Gas
Atoms in a gas zip past each other freely. The atoms in a gas will spread out to fill a container.

was allowing her a peek into the amazing world of physical science. What were the smallest pieces of matter? What happens to matter when it changes from solid to liquid to gas? How does electricity work? Sometimes just figuring out the right questions to ask was as exciting to Shirley as uncovering the answers.

~ "Can You Cook?"

After Shirley finished her freshman year, she decided to choose a career in physics. However, she soon found out that not all her teachers were as open-minded and welcoming as Professor French. Some of them still clung to their prejudices about women and African Americans. One of Shirley's professors approached her with a strange piece of career advice. "Colored girls," he said, "should learn a trade." *I am learning a trade,* Shirley thought. *I'm learning physics.* Professor John Wulff thought he was being helpful, but to a young college student he was hurtful.

Shirley was especially interested in learning more about materials science, the study of the properties of matter. She had taken a course from Professor Wulff and had done exceptionally well—far and above any of the other students in the class. When the semester was over, she approached him about hiring her for a summer job in the materials science lab on campus. There she could earn class credit as well as some pocket change. Though she didn't have any lab experience at that point, she thought she had a good chance of getting hired because she had done so well in his class.

Her interview was brief—and strange.

"Can you cook?" Professor Wulff asked.

"Excuse me?" Shirley replied.

"Can you cook?" he repeated. "Can you make anything?"

"Of course I can. My mother taught me to cook." *What does he want to know that for?* thought Shirley.

"Well, good, you're hired."

Since this was the same professor who had advised her to learn a trade, she thought: *Is Professor Wulff really asking me to be his cook?* "To do *what?*" she asked him, bracing herself for the answer.

The late Professor Wulff *(right)* is honored annually for his teaching skills with the Wulff Lecture. The lecture is given to freshman and sophomore students interested in materials science.

Sizing It Up

All things are made of tiny particles called atoms. The chair you sit in, the broccoli you eat, and the air that you breathe are all made of atoms. How small are they? In mathematical terms, the radius of an atom is 1×10^{-8} cm (known as an angstrom). That's one ten-millionth of a millimeter.

How do you picture something that small? You can't. But you can get a sense of its relative size, or how small it is compared to things you know. For example, think of how small an apple is compared to Earth. That's how small an atom is compared to an apple.

As tiny as an atom is, it is made up of even smaller particles. An atom contains a nucleus (core) made up of protons and neutrons. The protons and neutrons are held together by a force of nature called the strong force, which acts as a kind of glue. Around this core whirls a cloud of electrons.

An atom's nucleus is only one millionth of a billionth of the full volume of the atom. That means that the atom is mostly empty space. How empty? If you expanded an atom to the size of a football stadium, the nucleus would only be the size of a grape. (*This picture is not drawn to scale.*)

Also interesting is that this nucleus contains nearly all the atom's mass. That means that the grape would be almost as heavy as the football stadium!

"To work in the lab!" he replied, as though she were dim. "I figured if you could cook, you could work with your hands."

At first Shirley thought his question was odd. Later she realized it was fitting. Cooking involves the ability to follow a recipe, plus a little creativity—both of which are very useful in the lab.

~ Crystal Palace

The MIT Sloan laboratory was nothing like the simple chemistry lab back at Shirley's high school. While the high school lab contained only some sinks, Bunsen burners, and beakers, the MIT lab was a huge place with many different rooms. There were giant furnaces for melting samples, lathes to cut metal, and machines

The MIT Sloan Labs where Shirley worked were huge. Giant furnaces *(right)* were used for melting samples. Students place materials in receptacles for heating in the metal-lurgical labs *(below).*

with big rollers to flatten materials. It was like a colossal version of her father's workshop, with tools and machines everywhere.

Shirley's first assignment in the lab was to investigate an unusual occurrence known as "tunneling in superconductors." Tunneling refers to how electrons move through materials when there are barriers to their motion. Super-conductors are materials that, at ultra-cold temperatures, allow electrical current to flow very efficiently. If Shirley's work could help find ways to use electrical energy more efficiently, then more cost-effective electrical devices (ones that would operate at ultra-low temperatures) could be built. *(See box, page 37.)*

Making a Crystal Conductor

For her experiments with superconductors, Shirley needed to create a single crystal of a special metallic material. This material would be an alloy (or mixture) of niobium and titanium metals.

At first, the niobium-titanium metal alloy is in a multicrystalline form. The atoms are arranged in small crystals, like grains of sand all stuck together. Shirley needed to change the alloy from multicrystalline to a single crystal. (In a single crystal all the atoms are lined up in a set of rows and columns.) Shirley heated the alloy, turning it into a liquid. Then she cooled it very slowly, causing the alloy to become one large and well-ordered single crystal.

Next, Shirley performed tests to see if the alloy could be used to study tunneling of an electrical current. She exposed the alloy to oxygen. The metal atoms on the surface of the alloy crystal combined with the oxygen to form a metal oxide—a chemical compound, like rust. This very thin oxide coating became an insulator for Shirley's tunneling experiments.

She then put this metal and oxide wafer inside a vacuum-sealed gas jar along with a piece of another rare metal called indium. She heated the indium until it vaporized and condensed into a thin layer on top of the oxide. Both the indium metal and the niobium-titanium metal alloy were electrical conductors, and now they were separated by the insulating oxide layer.

Finally, Shirley attached very thin gold wires to the two metal layers so that she could measure the current flowing through the material "sandwich." When current was passed into the indium at very low temperatures, it could be measured on the niobium side. This showed that tunneling was occurring— electrons were passing through the insulating oxide barrier.

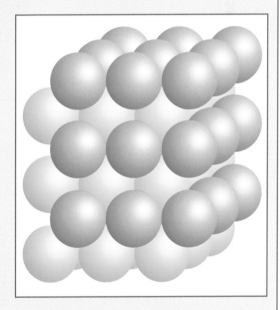

In the niobium-titanium alloy Shirley created, niobium atoms (yellow) and titanium atoms (green) formed a structure known as a body-centered crystal lattice. In this structure the atoms stack together much the same way that you might stack oranges and apples if you wanted to neatly mix them together.

Shirley's work in the lab required both physical and mental skill. Many of her experiments involved working with high heat sources on tiny bits of material. She often had to wear goggles to protect her eyes and gloves to protect her hands. Then there was the careful recording of data and the analysis of her results. Her long hours in the lab were both exhausting and exhilarating. Shirley learned a great deal. She also became convinced that the world of physics was the right place for her to be.

~ Choices, Choices

When Shirley, shown here in her official college graduation photo, was in her senior year at MIT, she had her eyes on the future. Graduate school was the plan, but where would she go?

During her senior year at MIT Shirley had to write a senior thesis (a long research paper). Her thesis would explain the results of the experiments she had been working on. In writing her paper, Shirley drew on a special mathematical formulation known as the BCS Theory. Basically, this theory helped explain superconductivity—or why some materials are better than others at allowing electricity to pass through at very low temperatures. The BCS Theory got its name from the initials of the three physicists who came up with the theory: John Bardeen, Leon Cooper, and Robert Schrieffer. All three had won the Nobel Prize in Physics in 1972 and had become role models for other physicists.

At the end of her senior year, when it came time to apply to graduate school, Shirley decided she wanted to learn from these physicists in person. She applied to the University of Pennsylvania because Schrieffer taught there and to Brown University because Cooper taught there. She also applied to the University of Chicago because it was close to the University of Illinois, where Bardeen taught. (She wanted to avoid the University of Illinois because she had been told it was not very welcoming to African Americans.) For good measure, she also applied to Harvard and to MIT.

Shirley's decision about which schools to apply to was fairly easy. To her delight, all the schools accepted her. Now came the hard part: deciding which one to attend.

Instead of dwelling on the **hurt**,
she focused on
what she could do

to make things
better for other
African Americans at MIT.

5

SPREADING THE DREAM

I n April 1968, during her senior year at MIT, Shirley visited the University of Pennsylvania in Philadelphia. The school had invited her to come to the campus and to meet the research staff. Shirley accepted the invitation, hoping it would help her decide which graduate school to attend. One of Shirley's sorority sisters from the Pennsylvania chapter of Delta Sigma Theta offered to drive her to the airport after her meeting.

That evening, as the two women drove to the airport, they listened to the car radio. Suddenly, the broadcast was interrupted with a special bulletin: Dr. Martin Luther King, Jr., had been shot in Memphis, Tennessee. A short while later, a second bulletin announced that Dr. King had died. They were so shocked that they nearly drove the car off the road.

Martin Luther King, Jr., had been a heroic figure to Shirley for most of her life. When Shirley was in high school, Dr. King led 250,000 people in a peaceful march on Washington. He gave an inspiring speech called "I Have a Dream." Dr. King's dream was that if one worked hard and lived right, one would be judged by the content of one's character and not by the color of one's skin.

Shirley *(opposite)* decided to stay at MIT for graduate studies and work toward making the "dream" of Dr. Martin Luther King, Jr., *(shown above)* a reality at MIT.

Civil rights leader Martin Luther King, Jr., was assassinated in Memphis, Tennessee. Large crowds turned out for his funeral procession on April 9, 1968.

Shirley took his dream to heart. She admired the way Dr. King had sacrificed his own comforts to devote his life to ensuring civil rights for African Americans.

~ A Clear Decision

By the time Shirley returned to her dorm, her mind was made up. She would stay at MIT and attend graduate school there. Why the sudden change of heart? Shirley knew from her own experience that MIT needed to grow with the times and be more welcoming of minorities. She also knew that if she drew upon her own leadership abilities, she could be a strong force for change.

There was only one hitch. Until now, Shirley's interest in physics was in solid state matter (what is now called condensed matter physics). But MIT's strength was in nuclear and high energy physics, referred to as elementary particle physics. The two studies were very different.

Condensed matter physics deals with the traits of physical material and how it behaves. High energy physics looks deeper into the structure of matter to see how the smallest particles influence each other under different circumstances. Now that Shirley was committed to

> The theoretical physicist's "lab" is her mind, and her tools are pencil, paper, and—nowadays—the computer.

staying at MIT, she had to change her focus to high energy physics. Luckily, she found the switch fairly easy, especially since high energy physics was more mathematical in nature.

~ Thought Experiments

Shirley had one more decision to make. Would she become an experimental physicist or a theoretical physicist? Experimental physicists do their work mostly in a laboratory. They perform carefully designed experiments to try to learn more about the world. They then compare their results to what they expected the experiments to show. This is a "hands-on" science.

However, it's the theoretical physicist who first makes predictions about what should happen in an experiment. The theoretical physicist's "lab" is her mind, and her

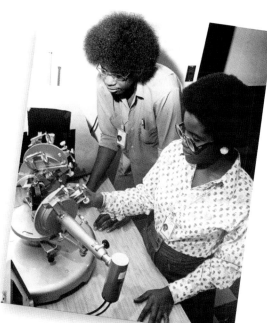

tools are pencil, paper, and—nowadays—the computer. The theoretical physicist conducts "thought experiments," starting with what is already known about the way the world works. She then asks "what if?" questions to frame a problem she wants to solve. Next, she puts the problem into mathematical form and solves it. The physicist can then use that result to explain the observed phenomenon. She can also make educated guesses about what might happen in different circumstances.

Experimental physicists, like the students above, conduct experiments in a laboratory. Shirley was a theoretical physicist conducting "experiments" in her mind.

For Shirley, the choice was clear: Theoretical physics meant using the mathematics she loved. But even more than that, theoretical physics would allow her to use her mind to create problems to solve, and then solve them. (*See box.*)

~ Forming a Union

When Shirley began her graduate studies in the fall of 1968, her first problem had nothing to do with physics. Instead it had everything to do with why she had decided to stay at MIT—the lack of minority students and the poor treatment of those few that did attend the school.

Solving a Scientific Problem

As a teacher, Shirley focused on the process of problem-solving and the fundamentals of physics. She explains that solving scientific problems involves a seven-step process:

1. Observe something happening or gather together information from previous experiments.

2. Make some assumptions (what you believe to be true) about what's happening based on existing theories.

3. Ask yourself how you can prove if those assumptions are true or false. Or ask how you can use those assumptions to make predictions about something else that can be tested.

4. Take the resulting word problem and convert it into a mathematical problem.

5. Solve the problem. Figure out the math and come up with a numerical answer as a prediction for a new experiment. The answer you get may require creating new theories.

6. Interpret your result. Figure out what it means and whether it makes sense.

7. Tell people about the results and propose new experiments to test the theory.

Shirley's main goals were to increase the number of minority students at the school and to ensure that they were treated fairly. To do this, she brought together a small group of black students and formed the Black Students Union. She hoped this group would put pressure on the school administration to make these changes.

Shirley insisted that the Black Students Union take a scientific approach to their social work. She didn't want the group to just say that something was unfair. They had to gather data to back up their claims. She always asked the "what if?" questions and thought about possibilities others had missed.

During meetings of the Task Force on Educational Opportunity, Shirley (left) and Dr. Paul Gray (below) sometimes disagreed on issues, but they always showed respect for each other.

The group met in various places on the MIT campus and hashed out their grievances. They generated a list of 10 demands. They asked the administration to recruit more black students and make financial aid available to them. Also on the list were demands for campus policies to be friendlier to black students, for more black teachers and administrators to be hired, and for the establishment of a transition program to help incoming black freshmen adjust. The overall objective was clear: Black and other minority students must be better integrated into life at MIT.

The Black Students Union took their demands, or "proposals," to the associate provost, Paul Gray. (Dr. Gray would later become president of MIT.) He listened carefully to Shirley's thoughtful, reasoned arguments. As a result of that meeting, he formed the Task Force on Educational Opportunity and appointed Shirley as the student leader. The group included six faculty members and several administrators, along with six African American students.

Members of the Black Students Union stand proudly with their basketball tournament trophy.

The task force met every week, sometimes several times a week, often for hours at a time. They discussed the issues and developed policies and processes to help minority students. The meetings were often tense, but never confrontational. Shirley had strong positions and she spoke eloquently to defend them. She often disagreed with Dr. Gray on issues, but they never doubted each other's honesty and they always showed respect for each other. Shirley had a special talent for finding points of agreement on difficult issues. She was only 22 years old, but she often showed more sense and good judgment than all the others in the room.

Recruitment of African American students was a big focus of the task force.

~ On the Road

The task force accomplished much its first year. The school recruited black students through mailings, created a financial aid policy for minority students, and hired a number of black administrators and faculty. It also developed a summer program to help new minority freshmen prepare for MIT's rigorous curriculum.

In its second year, the task force expanded recruitment efforts by sending MIT students out into high schools. Shirley traveled to the Midwest, visiting such cities as Cleveland, Akron, Detroit,

46

ment efforts by sending MIT students out into high schools. Shirley traveled to the Midwest, visiting such cities as Cleveland, Akron, Detroit, and Cincinnati. She met with guidance counselors, principals, and assistant principals. She talked to top-ranked students about MIT and why they should apply. All together, Shirley and several other students on the task force visited more than 100 schools. The recruiting efforts paid off. In the first year after the recruitment program began, the number of entering African American freshmen at MIT jumped from 5 to 57.

The success of the Task Force on Educational Opportunity at MIT is reflected in the university's African American gospel choir group, shown here in the 1970s.

Shirley did all this traveling while still carrying a full course load. Fortunately, many of the beginning graduate courses covered much of the same material she had studied as an undergraduate student. Still, she eventually felt the strain of all the travel and cut back on her activities after a year. She needed time to study for the general exam, which all doctoral students have to take to qualify for their Ph.D. Though the exam must be taken by the third year, Shirley took it after attending graduate school for only a year and a half. And she passed!

~ Taking Dad's Advice

When Shirley was growing up, her father often repeated a bit of advice to her when she ran into difficulties: "You can't control everybody else, and you can't control the world and everything that happens, but you can have your greatest control—and certainly your greatest influence—by how you control yourself."

Shirley had taken her father's advice to heart and was living it to its fullest. She used the struggles she endured as an undergraduate as the catalyst for positive change. Instead of dwelling on the hurt, she focused on what she could do to make things better for other African Americans at MIT. She was truly carrying on the work of

Shirley was put in *charge*
of the whole physics program for
Project Interphase,

even though she was
still a **student** herself.

THE BIG PICTURE

6

The Task Force on Educational Opportunity made many changes at MIT. One of its most successful accomplishments was the creation of Project Interphase. This six-week summer program—named for the phase between high school and college—helped entering minority students prepare for a full course load in the fall. It gave Shirley an opportunity to share her knowledge with others and help them become successful MIT students.

The program included courses in physics, precalculus, and writing as well as some laboratory work. Shirley taught physics. Though she worked under the supervision of a professor, she had her own ideas about how the course should be taught. These ideas did not always mesh with those of the other faculty members.

At that time, the program was taking an experimental approach to teaching physics to make it more interesting to students. The professor in charge believed that it was important to get the students excited about physics. By working on different projects, they could see what learning physics would allow them to do.

Minority students *(above)* participating in Project Interphase benefited from the help of student instructors like Shirley *(opposite)*. To prepare new students for the challenges of MIT science courses, Shirley focused on the fundamentals of physics and problem-solving.

However, Shirley knew that making physics fun was not necessarily what these students needed. First of all, she knew that they were already motivated. They had chosen to come to MIT because they were already excited about science and math. Besides, to be accepted at MIT, they had to be top students. But Shirley had been through MIT's tough first year and she knew the coursework these students would face. The classes were extremely fast-paced and, unlike Shirley, many of the entering freshmen had not had the chance to study advanced science in high school.

If they were going to survive their first year at MIT, Shirley felt these students would have to have a firm grip on fundamentals. To ensure their success as much as possible, Shirley stressed working on problems, which she believed would give her students the foundation they needed to then work on projects. It was essential for these students to understand the big picture of what physics was all about. It was also important for them to be able to solve real problems. So she taught her classes with these goals in mind.

Project Interphase instructors demonstrated key concepts to reinforce basic scientific lessons.

Shirley concentrated mostly on the study of Newtonian mechanics, which has to do with the laws of motion and force. Newtonian mechanics is very important because it forms a basis for being able to analyze—through mathematics—every situation involving motion, ranging from making a better athletic shoe to designing a safer car to fighting fires to exploring space. Plus, students couldn't even begin to understand Einstein's theory of relativity until they understood Newtonian mechanics.

Shirley's approach of emphasizing the nuts-and-bolts fundamentals proved a great success. Some of the students in the more experimental classes switched to hers.

The following summer the program expanded to include more students, not just minority students. Shirley was put in charge of the whole physics program for Project Interphase, even though she was still a student herself. Being in charge forced her to organize the program and think about the most important concepts for new physics students. For instance, she felt they needed a firm grasp of

It's All Relative

In 1916, Albert Einstein published a theory that changed what everybody had thought about the way the world worked. He called it the general theory of relativity.

Einstein's theory proposed that gravity is a bending of space and time, and that this bending causes what we perceive as a force between bodies. Einstein said that space curves according to the bodies in it. Think of the way a trampoline curves when you place a heavy ball on it. This curvature causes bodies to "fall" toward other bodies, the way a marble would roll toward a bowling ball on a curved trampoline. Because space is curved, traveling light also "falls" as it nears a heavy body. Thus the path of light would bend. This phenomenon affects how we see the universe around us.

Einstein's theory was put to the test during an eclipse of the Sun in 1919. A distant star appeared to be in a different position from where astronomers knew it actually was. That's because its light was bent by the pull of the Sun. Thus, Einstein's theory was accepted as scientifically accurate.

Project Interphase instructor Brian Schwartz (right) conducts an experiment for new students.

mathematics in order to do physics problems. She also made sure they were able to do well in timed testing, which was a fundamental part of MIT's curriculum.

~ Doing for Others

In her last two years as a graduate student, Shirley lived in an apartment off-campus. It was a one-bedroom apartment on the first floor of a two-story house. She furnished it simply but comfortably with some beanbag chairs and lamps in the living room; a small table and chairs in the dining room; and a bed, dresser, and chest of drawers in the bedroom.

Her apartment was also filled with piles of papers—old problem sets and exams—and, in her words, "a gazillion books." She bought as many books as her finances would allow—not just required textbooks, but any book she could find that related to the subjects she was studying. Thus outfitted, her apartment now enabled her to do for other African American students what had not been done for her: invite them to partici-pate in study groups. At Shirley's apartment students could study old exams and use her

> Everyone was very serious about their work, and they all had a strong determination to do as well as they could at MIT.

physics study sets, which she had completed two or three years previously. The students could also use her impressive library of books.

Shirley's apartment became a central gathering point for the African Americans just starting out at MIT, most of whom were still men. Everyone was very serious about their work, and they all had a strong determination to do as well as they could at MIT.

One student in particular spent a great deal of time studying at Shirley's apartment. His name was Ron McNair. When Shirley first met him, in 1970, he was an undergraduate student at North Carolina Agricultural and Technical State University and was spending the spring semester of his junior year studying at MIT. This was part of MIT's effort to recruit outstanding black students from other colleges. Ron did indeed return as a graduate student in 1971, and that's when he and Shirley became good friends.

Ron often spent hours studying and reviewing Shirley's old problem sets in the living room of her apartment, while Shirley did her calculations for her graduate thesis in the dining room. Periodically, Ron would politely interrupt Shirley's work

Shirley befriended Ron McNair, a graduate student doing work in laser physics. His strong determination to succeed despite obstacles reminded Shirley of herself.

with questions about what he was working on, and she'd go over the material with him.

One night Ron was mugged in Cambridge. His stolen briefcase contained all of the data he had collected for his thesis. He had spent years doing experimental work in laser physics. Now all of that data was lost.

With Shirley's moral support and encouragement, and his own determination, Ron worked day and night to reconstruct his experiments. He reproduced his work—and finished his thesis— in a little under a year. Ron's determination not to allow life's circumstances to get in the way of his goals reminded Shirley of herself. Like her, he had a strong inner force that kept him pursuing his dream. Though they each pursued different careers, they would remain friends over the years.

~ Stung!

Shirley's third summer of graduate school, in 1971, took her to Boulder, Colorado. There she enrolled in a prestigious annual theoretical physics summer school at the University of Colorado. She attended a series of lectures on various topics given by physicists and professors from all over the country. Two of those speakers would have a big impact on Shirley.

The first was Harry Morrison, a professor of theoretical physics at the University of California at Berkeley. It was the first time she had met an African American theoretical physicist who was a tenured professor at a major university. She became good friends with Dr. Morrison and his wife, Harriet, and often visited their home. Shirley enjoyed knowing someone as intrigued by knowledge as she was. And the books! Like Shirley, he had stacks and stacks of every kind of book on physics and mathematics.

The other lecturer was a theoretical physicist from AT&T Bell Laboratories named John Klauder. He would later play a small but vital role in Shirley's career.

The math and physics lectures energized Shirley's mind. But she was physically energized by hiking. She often went hiking in the

Colorado Rockies. It was a memorable summer, made even more memorable by something Shirley learned about herself and bees.

One day Shirley was walking past a shrub when she suddenly felt a stabbing pain in her right pinky finger. She looked at her pinky, and to her utter amazement, there was a bee hanging on it! She flicked the bee away, but the stinger remained. After she pulled it out, her hand began to swell, and it swelled right up to her wrist. She sought medical attention immediately and learned that she was allergic to bee venom. She thought back to her summers collecting bees in jars and all the times she had handled them, never once getting stung. She had no way of knowing how sensitive her allergy might have been as a child, but she still felt lucky she had never had to find out.

> For her father, seeing Shirley receive her Ph.D. was the realization of all his hopes and dreams.

~ First of Many Firsts

In the fall of 1973, Shirley received her Ph.D. from MIT. She was the first African American woman to receive a Ph.D. in any field from the school.

The annual commencement ceremony for the graduates of MIT took place in June 1974. The ceremony was held in the Rockwell Cage, a World War II surplus airplane hangar that was now used as an athletic facility. As supportive as ever, Shirley's parents and her sister Gloria were in the audience when the ceremonial Ph.D. hood was placed on Shirley. Even her math teacher from high school, Mrs. Smith, was there. For her father, seeing Shirley receive her Ph.D. was the realization of all his hopes and dreams. His daughter was now Dr. Shirley Ann Jackson.

For Shirley, it was just another step in a lifetime of firsts.

Shirley was even
more **impressed**
by what lay below.

A four-mile ring
buried beneath the land
circled the facilities.

HIGH ENERGY DAYS

7

S hirley had always been a city girl. She was raised in
Washington, D.C., and went to school in Boston. But right
after earning her Ph.D. in 1973, she found herself on the
wide-open prairie of the Midwest. Shirley was working in a postdoctoral (continuing study) position in the theory group at the
Fermi National Accelerator Laboratory (now called Fermilab) in
Batavia, Illinois. The lab was located on 6,800 acres of farmland
35 miles west of Chicago. It was devoted to research on subatomic
particles—the tiniest pieces of matter known—and their interactions.

When Shirley began her postdoctoral work at Fermilab, the graceful skyscraper known as Wilson Hall *(opposite)* was not yet completed. But bison *(above)* roamed the campus—and still do.

Shirley was amazed at the expanse of land on which she now
worked. The landscape was dotted with one-story houses and
buildings called barracks where the scientists worked on various
experiments. Even more astonishing was the herd of American
bison grazing on the land around the laboratory. Robert R. Wilson,
the founding director of the lab, had brought in the bison as a
reminder of Illinois's prairie history and to serve as a symbol that
the lab was on the frontier of high energy physics.

Shirley was even more impressed by what lay below. A four-mile
ring buried beneath the land circled the facilities. This was the
famed proton accelerator ring, where the physicists sent particles
zooming at nearly the speed of light on a path of collision.

~ Policy vs. Practice

Although Fermilab director and founder, Robert Wilson, ensured a policy of human rights at the lab, not all of the scientists were ready to fully accept an African American woman engaged in research.

Fermilab director Robert Wilson actively promoted a human rights policy. He had posted a formal and specific policy to ensure that everyone at the lab "can live and work with pride and dignity without regard to such differences as race, religion, sex or national origin." The statement, which Wilson had originally sent to Dr. Martin Luther King, Jr., also said, "In any conflict between technical expediency and human rights, we will stand on the side of human rights."

Still, others in the physics community were not yet ready to fully accept Wilson's idealism. They were not accustomed to African American women engaged in important research and did not know what to make of Shirley. Though they were not outwardly hostile, their lack of friendliness reminded her of her undergraduate years at MIT.

The world beyond Fermilab was also less than welcoming. Initially Shirley stayed at a hotel fairly close to the lab. But after nearly 30 days there, she felt it was time to find a place to live. She tried to find an apartment close to the facility, but landlords did not want to rent to black tenants. Though great strides had been made through the civil rights movement, many people still actively discriminated against African Americans.

Shirley kept searching farther from the facility, but she still had no luck. Once, she found a newly refurbished apartment she liked. But when she was ready to put down a deposit, the landlady said, "I've got to talk to my husband first."

The next day, Shirley called about the apartment, eager to finalize the deal. But the woman said, "I'm sorry, Ms. Jackson. You seem like a very intelligent young lady, but my husband and I had trouble before when we rented to a Hispanic doctor. We're not going to rent to you."

That was the last straw. Shirley had no choice but to move to Chicago. She would be much farther away from the lab than she wanted to be, but at least the city was more open to renting to African Americans.

Shirley did find an apartment in the Windy City, but now she had a long commute to get to her job. To cut down on the cost of driving herself, Shirley relied on a car-pool system. She shared the 70-mile round-trip with three computer programmers who worked at the lab. They all lived in different parts of Chicago, so they would meet each morning at a place called Soldier Field, near the downtown area. The car-pooling worked well as long as Shirley kept the same hours as her car mates. If she wanted to work late it was a different story. Then she had to find someone to take her to the train station, take the train to Chicago, and switch trains to Soldier Field to finally get her car and drive home.

Looking back at her time at Fermilab, Shirley *(above)* fondly remembers trudging through the snow discussing physics with colleague Mary K. Gaillard. She enjoyed the beauty of the natural surroundings *(below)* of Batavia, Illinois.

~ Discovering Chicago

Though Chicago had not been her first choice, Shirley soon found that the city had its benefits. She discovered Chicago's museums and nightlife, and made a wider circle of friends than she probably would have made if she had lived closer to the lab.

Gradually, the other researchers' coolness toward Shirley began to thaw as they realized how smart and serious she was. Shirley began making friends. Many days, she shared her long walk between the barracks and the cafeteria with Mary K. Gaillard, a visiting scientist from Europe. On these walks, often trudging through the snow, Mary and Shirley struck up

a friendship. They swapped ideas about politics and physics and life in general. Though they had different personalties, they held many common views about what was important to them.

~ Off to Europe

After a year at Fermi, Shirley applied to the Ford Foundation requesting a postdoctoral fellowship (a grant to help with expenses) to work at the European Organization for Nuclear Research (CERN)

Smashing Particles

Subatomic particles are so tiny they cannot be seen even through the most powerful microscope. How can you study things that can't be seen?

One method is to use a particle accelerator *(below)*. These huge machines—sometimes miles long—propel particles at nearly the speed of light and then smash them into a piece of matter or each other. Millions and millions of other collisions result—about 2 million a second—as the particles ricochet off of each other.

The collisions force the particles to scatter into fragments, called subatomic particles.

Researchers don't actually see the fragments. Instead, they use sophisticated detectors, some as large as a small apartment building, to see the "signatures" of the collisions, such as light tracks or energy bursts. Computers process the information from the detector and display the tracks on a screen.

in Geneva, Switzerland. This is one of the world's largest particle physics centers, where thousands of physicists from all over the world come to work. Since its creation in 1954, CERN has made many important discoveries and contributions, including the World Wide Web. The Web was originally developed as a way to improve and speed up information sharing among physicists working in different universities and institutes all over the world. Now it's hard to imagine everyday life without it!

Because Shirley had had a Ford Fellowship as a graduate student, the Foundation was familiar with her and her work. So although

The researchers then "read" all this information to learn about the particles and the forces that control their interactions. These tracks can help scientists figure out the mass and electric charge of the particles that made them.

The photo at left shows only a small segment of the particle accelerator at CERN. The light tracks, or signatures, left behind by a particle collision produce computer images *(above)* that unlock the secrets of their interactions.

In this aerial view of CERN you can see the Alps in the background. Shirley often went skiing and hiking in these majestic mountains.

the foundation did not ordinarily grant postdoctoral fellowships, they offered Shirley an individual grant, which provided a stipend and travel money. CERN offered to supplement the grant to help Shirley pay for her living expenses, which were much higher in Switzerland than in Chicago.

Shirley jumped at the opportunity to go to CERN, especially because it gave her the chance to work with her friend Mary, who had returned there after her position at Fermilab ended. Together they collaborated on a paper on neutrino experiments. Neutrinos are the most mysterious of the known particles in the universe. They are electrically neutral and have virtually no mass. Like ghosts, they can pass through any solid material, no matter how thick or dense, and emerge virtually unchanged.

Even though Shirley had never lived outside of the United States, she was an experienced traveler and quickly adapted to Swiss life. Swiss francs replaced dollars in her purse. Italian and German words began to creep into her vocabulary. She studied French. Snow-covered mountain peaks and centuries-old buildings soon became a common sight for Shirley.

Every Sunday she visited a restaurant in the old part of Geneva, where she ordered cheese fondue, a Swiss dish in which chunks of French bread are dipped into melted cheese. There, she read the French newspaper *La Monde* and the *International Herald Tribune*. Afterwards, she would wander over to a nearby park and read a novel.

Best of all, Shirley loved working at CERN, where she finally found the companionship that had been lacking at Fermilab. The scientists who worked at CERN came from all over the world, so they were more tolerant of people of different races and ethnicities. Also, the European and Asian countries they came from did not have the history of slavery and other race issues that the United States did.

Shirley found it easy to make friends with her colleagues, and she thrived in this highly social environment. Her work friends had her over to their homes for dinner, and she in turn made dinner for them. She also learned how to ski and went hiking in the Alps. She especially enjoyed her friendship with Barbara and Fridger Schremp, a German couple who were both physicists at CERN. Often she would drive with them up into the mountains. On a grassy spot overlooking Geneva, they would have a picnic

Geneva has the second-highest population in Switzerland (after Zurich). Seen in the distance is one of the city's most visible landmarks, the Jet d'Eau fountain in Lake Geneva. It shoots a 460-foot column of water into the air.

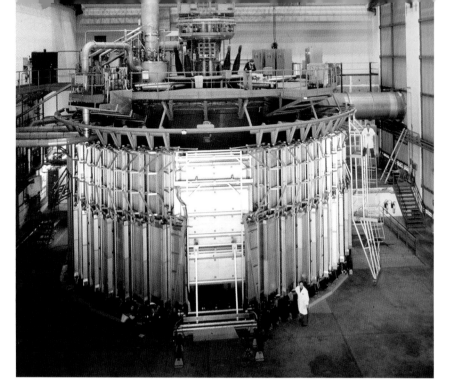

In addition to the particle accelerator, CERN's facilities included the Big European Bubble Chamber, one of the largest particle detectors in the world.

lunch of lamb grilled on a hibachi, along with wine and cheese. As they ate, they listened to Tina Turner's soulful singing on a portable tape player.

~ The Charm Quark

Samuel C. C. Ting, a professor Shirley knew from MIT, had a research group at CERN and often visited there. In 1976, he received a Nobel Prize for the discovery of a new particle, the J/psi.

By chance, Shirley happened to be at CERN at a very exciting time in the field of physics. In November 1974, a new particle was discovered, and one of the researchers who discovered it was Samuel C. C. Ting, a professor Shirley had known at MIT. (The other researcher was Burton Richter of the Stanford Linear Accelerator Center near San Francisco.) Though Ting was now at Brookhaven National Laboratory on Long Island, he had a research group at CERN and often visited there. Ting and Richter independently discovered the J/psi particle, which is made of a charm quark paired with an anti-charm quark. (An anti-charm quark has the same mass as a charm quark but is opposite in all its other properties. For example, it has an opposite electric charge.)

This discovery was momentous, and it won Ting and Richter a Nobel Prize in 1976. Events like this one were exciting to scientists such as Shirley. They showed that the great ideas that theoretical physicists described could often be proven to be scientifically accurate. This made the long hours and years of research seem more worthwhile. It also inspired scientists like Shirley to keep searching for the right questions to ask and answer.

After her yearlong fellowship at CERN ended, Shirley returned to Fermilab, where she completed the second year of her post-doctoral position. But she now had a big decision to make. Would she continue in high energy physics, where jobs were particularly hard to come by, or would she return to her original field of condensed matter physics, where there were more opportunities?

It was a dinner meeting at a physics conference in Atlanta that set Shirley on the next path in her career.

As she practiced her **speech**,
her excitement about

the upcoming **challenge** grew.

8

THE BELL YEARS

I n March 1976, Shirley was in a restaurant in Atlanta having
dinner with Maurice Rice, the head of AT&T Bell Laboratories.
Founded in 1925, Bell Labs invented and developed some of the
most important communication technologies in use today. Devices
like stereo recording, sound motion pictures, the touch-tone phone,
fax machines, communications satellites, lasers, and cell-phone
technology all got their start at this research company.

Shirley had come to Atlanta for a meeting of the American
Physical Society, an organization of thousands of physicists.
Its purpose is to increase and spread the knowledge of physics.
Shirley had been a member since her graduate school days. Her
meeting with Dr. Rice had been set up by John Klauder, the Bell
Labs theoretical physicist Shirley had met in Boulder, Colorado.

Shirley told Dr. Rice of her interest in switching fields from high
energy physics to condensed matter physics, with which Bell Labs
was involved. She talked to him about her work with neutrinos at
CERN. He was very quiet during their meeting, and Shirley had a
hard time figuring out what he was thinking. But after 45 minutes,
he invited her to come to Bell Labs in New Jersey and give a seminar.
Clearly, he was impressed by her.

Like all the theoretical researchers at Bell Labs, Shirley used a blackboard to work out problems. Above is a portion of a mathematical equation that Shirley worked on while at the technology company.

Shirley knew the seminar would be her job interview. Physicists at Bell Labs were expected to communicate well. A large part of their job involved traveling around to give talks explaining their work to other scientists. Since Dr. Rice had been impressed with her work on neutrino physics, she decided she would talk about that.

Shirley knew a lot was riding on her presentation. Her mind was filled with questions. Would she be able to explain herself clearly? Would the audience be both interested and informed by what she had to say? But she didn't allow these thoughts to make

Bell Labs believed that it wasn't just the discovery of knowledge, but also the sharing of it that helps to advance scientific innovation.

her nervous. Winning the oratorical contest in high school had helped her recognize that doubts like these are natural before giving a speech. And her understanding of the subject matter was thorough and well researched. As she practiced her speech, her excitement about the upcoming challenge grew.

Shirley gave her lecture to a roomful of physicists—both theoretical and experimental—in a large conference room at Bell Labs in Murray Hill, New Jersey. Not surprisingly, she did very well with the talk. But the interview process was not over. She now had to do a series of one-on-one conversations with researchers in the theoretical physics group. Shirley wanted to show that she could make the transition from her present field of high energy physics to condensed matter physics. So she explained how her theoretical work in neutrinos could be adapted to help advance their work in condensed matter physics.

~ Dream Job

Shirley's intelligence and creativity won her a one-year job as a limited-term member of the technical staff, even though she had not had any work experience in condensed matter physics since her undergraduate days. Imagine being a tennis player and then suddenly changing to swimming. They're both sports, but they call on completely different skills and training. That's what it was like for Shirley to change from high energy physics to condensed matter physics. But she did so well that when the year was up, she was made a full-term member of the technical staff.

Shirley was not expected to help AT&T develop new devices or technologies or even improve existing ones. That was left to the scientists who worked in the applied research department. Instead, Shirley and the other theoretical physicists were expected to discover important problems to work on and then write papers that would be published in the major scientific journals. She was also expected to speak about her research all over the world. Bell Labs believed that it wasn't just the discovery of knowledge, but also the sharing of it that helps to advance scientific innovation.

Bell Labs believed that no one could tell where the next great breakthrough would come from. The way they saw it, if they told their researchers what they should be looking for or working on, that could limit their creativity. For instance, when transistors were invented in 1947, the scientific community originally thought they would be used in hearing aids. It turned out that they could do much more. Without transistors, the world today would be a different place. They are critical components in everything from televisions and cell phones to cars, airplanes, and computers.

Telstar, a communications satellite created by Bell Labs, transmitted the first TV picture—an American flag—from space on July 10, 1962. It went on to transmit international phone calls, television programs, radio signals, and newspaper stories.

~ Organized Chaos

Because she was a theoretical physicist, Shirley did not work in a laboratory, but in an office. Her office was filled with books and journals, pads of paper and pencils, a computer, and of course, a blackboard covered with mathematical computations. Anyone stopping to visit her might be surprised at the mess. Papers were everywhere! But Shirley knew exactly where everything was. Like many of the other researchers, she thrived in a kind of organized chaos.

At Bell Labs, Shirley Jackson was making a name for herself in theoretical physics by coming up with new theories about the behavior of electrons.

Shirley used four basic tools to do her work: a pencil, a pad of paper, a computer, and her imagination. She spent hours reading, thinking, and learning new mathematical techniques to apply to a particular problem. She also reviewed other scientists' calculations and examined how they cast their problems in mathematical form, then did it herself. But she wasn't just doing mathematics. A theoretical physicist uses mathematics as a tool to get the answer to a physical problem or to describe a physical situation.

Much of Shirley's work had to do with semiconductors. These are materials that conduct electricity only under certain circumstances and are used to control current in electronic equipment. Because semiconductors don't exist in nature, they must be "grown" in a machine by combining very thin layers of materials such as silicon, arsenic, and phosphorus. The problem is that when such materials are layered, sandwich-like, they pull each other in different directions. That creates stress and strain that can lead to defects and may make the material useless.

Shirley thought about this problem for a while. As she went over the issues again and again, she gradually came up with some new ideas. Finally, Shirley had a breakthrough. She developed a theory of how to predict the strain that would happen when these materials were layered. When the amount of strain was known, scientists could tell how thick the layers could be grown before becoming defective.

Shirley's work was extremely important in making the semiconductor lasers that we now take for granted. Every time you listen to a CD, scan an item at the grocery store, or slide a DVD into your computer, you're using technology that Shirley's work helped improve.

> Papers were everywhere! But Shirley knew exactly where everything was. Like many of the other researchers, she thrived in a kind of organized chaos.

~ Adventures in Helium

Shirley also worked on another problem that helped her make a name for herself in theoretical physics. She wanted to know what happens to electrons when they are exposed to light and other things in two dimensions—that is, on the surface of a material.

Electrons behave very much like marbles. To get a sense of the way she worked on the problem, imagine rolling a marble around on a table that has some rough spots on its surface. You want to predict where that marble is going to end up. Because of the nicks and bumps in the table, you can't say for sure what path the marble will take or where it will end up. You can only predict the *probability* of it moving in certain directions.

By looking at all possible paths across the table, you can come up with a theory about the most likely way the marble will travel. You also might be able to predict how fast it will get across and where it will end up on the other side of the table.

If you start adding marbles, however, the problem gets even more interesting. Now you need to think about how they might run into each other and how that will change the probability of each marble getting across the table.

Instead of a rough tabletop, Shirley was looking at films of liquid helium that were put on different kinds of materials, or substrates, such as silicon. Helium is the gas that keeps balloons and blimps aloft, and it is one of the lightest elements in the universe. At super-cold temperatures, however, helium turns into a liquid. Shirley found that she could change the properties of a liquid helium film by changing the thickness of the material on which it was placed. This change then alters the way electrons behave on the film's surface. The interactions between electrons in the liquid helium film are very similar to the interactions electrons experience in semiconductor devices—the chips used in almost everything from your watch to your computer.

Shirley did not come up with her theory overnight. She worked on the problem for years, filling notebooks with her calculations and alternating with other work before she came up with a workable solution. That may seem like a long time to work on one problem, but it's the norm in the world of theoretical physics.

Shirley's hard work and tenacity paid off. She was able to make important new predictions about the behavior of electrons that were later confirmed by experiments. Because of her advances, she was elected a Fellow of the American Physical Society in 1986, a major milestone in any physicist's career. Being made a fellow to this prestigious society gave supporting evidence of her ability in physics. It also showed Dr. Rice's wisdom in hiring her and trusting that her drive, intelligence, and creativity would enable her to make the switch from high energy physics to condensed matter physics.

~ Tea and Cookies

Shirley's work routine entailed a lot more than sitting at a desk writing mathematical equations. Bell Labs understood that great scientific strides could be made when scientists were encouraged to share their ideas and knowledge with each other. Throughout the day, Shirley and her colleagues roamed around the offices to see what the others were working on. They visited the labs of the experimental researchers. They had journal clubs in which they took turns picking articles from scientific journals on a subject that interested them and doing a presentation and critique of the main ideas. Everyone was learning all the time.

Every day at about 3 or 4 P.M., work came to a standstill. This was teatime—another chance for the researchers to get out of their own minds and discuss their ideas and problems with others. They would wander into the tearoom and flip through the many journals kept there. They would also share their frustrations about problems they were having difficulty working out and their excitement at new solutions. All this productive "chitchatting" was accompanied by tea and cookies.

~ Sharing Knowledge

Travel was a big part of Shirley's work. Several times a year she attended conferences and gave talks. She traveled all over the United States. She also went to France, Belgium, Germany, Italy, Hong Kong, Japan, and the United Kingdom. Shirley often felt like she had hit the jackpot. Not only was she able to speak about her favorite subjects, but she was also able to see the world while doing it.

As she had done at MIT, Shirley mentored graduate students who were working at Bell Labs on fellowships. Most of these students were men because physics was still a field very few women pursued. One young man, Anthony Johnson, particularly captured Shirley's attention. He often stopped by her office for advice or to chat about how things worked at Bell Labs. He shared with Shirley what he wanted to do in his career. His biggest hope was that he would become a member of the staff at Bell Labs.

> Shirley's hard work and tenacity paid off. She was able to make important new predictions about the behavior of electrons that were later confirmed by experiments.

To help him achieve his goal, Shirley gave him pointers about his research and tips on how to interact with the other physicists. To Anthony, many of the physicists seemed like larger-than-life figures because of their reputations as world experts. But Shirley let Anthony know that these towering figures were every bit as human as he was. Her advice helped him feel as though someday he could be a member of "the club." And, eventually, Anthony did become a full member of the technical staff.

Anthony especially admired Shirley's fearlessness and tenacity. She knew what she wanted. She was the only African American woman doing theoretical physics at Bell Labs, yet that did not seem to hold her back. She had been able to gain respect because of her knowledge and talent.

~ The Best of Everything

Shirley loved everything about her job at Bell Labs. She loved thinking about a problem and figuring out how to solve it, then writing about it and getting it published. She enjoyed interacting with her colleagues and giving talks all over the world. She also appreciated having so many resources at her fingertips. The whole package was great fun for Shirley.

Shirley and her husband Morris Washington met at Bell Labs.

Her work at Bell Labs had a personal benefit as well. One day about a month after she began working there, she met a fellow physicist named Morris Washington at a seminar. The two noticed each other from across the room. Shirley thought Morris was attractive. After talking to him, she knew it was more than his looks that attracted her. The two clicked almost instantly. Within a few minutes, Shirley and Morris learned that they had been born just a few days apart.

Shirley and Morris had been raised very differently from each other. He grew up on a farm; she grew up in the city. He was an only child; she was one of four siblings. But they had two big things in common: their interest in physics and the close bonds they had with their families.

Shirley spent every holiday with her family and visited them often. And though Shirley and her sister Barbara lived on opposite sides of the country, they made a point of visiting each other at least twice a year.

At first, Shirley and Morris were friends. However, after awhile, their friendship grew into dating. Finally, four years after they met, Shirley and Morris were married. Their wedding was a joyous occasion. Members of both families helped celebrate their union. But the best was yet to come. A little more than a year later, in 1981, Shirley and Morris had a child. They named their son Alan.

Shirley stayed home and doted on Alan full-time for three months. But then she had to think about getting back to work, so a good day-care arrangement was needed. A woman Shirley had met through one of her sorority sisters had a one-year-old son who was being very well cared for by a woman who looked after a couple of children in her home. Shirley and Morris met her and decided they felt comfortable leaving Alan with her while they were at work.

Once Shirley returned to Bell Labs, she and Morris worked out their schedules so that they shared the responsibility of getting Alan to and from day care. Because Morris worked in experimental physics, it was usually harder for him to stop whatever he was

Shirley was a devoted mother *(above)*. In the photos below, the family celebrates Alan's first birthday. Shirley *(left, center)* is pictured with her mother *(left)* and her mother-in-law. Her father and Alan are shown below.

doing at a particular time of day. It's often not possible to stop an experiment in the middle and pick it up the next day. However, as a theoretical physicist, Shirley *could* stop at a certain time. So typically, Morris would take Alan to day care in the morning, and Shirley would pick him up in the afternoon. The whole family had dinner together as often as they could. Shirley and Morris worked hard to nurture a strong family life while growing in their careers.

~ Expanding Her Reach

By 1985, Shirley had been at Bell Labs for nine years. She enjoyed her work tremendously, but felt a need to expand her knowledge and expertise into the larger public sphere. So when New Jersey Governor Thomas Kean asked her to become a founding member of the New Jersey Commission on Science and Technology, she jumped at the opportunity.

The commission's purpose was to create partnerships between industry and the government. These partnerships were intended to spur research and investment in areas important to New Jersey's economy, such as advanced biotechnology and medicine, fiber-optic research, and hazardous substance research. Though it was an unpaid position, it was a huge honor. Shirley did this work while still holding a full-time position at Bell Labs.

Not everyone greeted Shirley's work on the commission enthusiastically. Some of her fellow researchers at Bell Labs felt that the requirements of her new position—traveling, speaking at events, serving on committees, advocating for public policy—were not things a scientist should do. No one pressured her to quit her policy work, but Shirley sensed their disapproval. However, as she often had in the past, Shirley relied on her own inner voice to tell her what was right. And that voice kept telling her that her policy work was important to continue.

Initially, Shirley was appointed to a four-year term. Her contributions were so valuable, however, that she ended up serving under three successive governors, for a total of 10 years.

~ Losing a Friend

On January 28, 1986, a colleague approached
Shirley and told her that the NASA space
shuttle *Challenger* had blown up minutes after
launching. Shirley's thoughts immediately
went to Ron McNair, a graduate student she
had mentored back at MIT. A veteran astronaut,
Ron had flown shuttle missions since 1984.
Shirley and Ron had remained friends over
the years, and they had spoken to each other
on the telephone just the week before.

After the space shuttle
Challenger blew up,
Shirley mourned the
loss of her friend, Ron
McNair *(front row on
right)*, whose courage
she had always admired.

With a sinking heart, Shirley realized the tragedy now before her.
Ron had been on the ill-fated *Challenger*. Shirley's heart ached as she
thought about Ron's two young children and how hard it would
be for them to lose their father. She remembered what a brave
person Ron had been and how hard he had worked throughout

his life. The whole world mourned the loss of the *Challenger's* crew. Shirley knew she would never forget her courageous friend Ron.

~ Back to School

In 1991, after 15 years at Bell Labs, Shirley felt that it was time for a change. She decided to return to academic life as a teacher. Mentoring students at Bell Labs had been very gratifying, and Shirley thought that teaching would allow her to have a greater influence on the future of physics by educating the next generation of physicists.

When administrators at Rutgers University in Piscataway, New Jersey, learned of Shirley's interest, they invited her to come to the school and give a seminar on theoretical physics. As usual, Shirley wowed her audience. In fact, the administrators were so impressed by her teaching skills that they offered her a job as a tenured full professor (the highest teaching and research position universities offer).

Shirley wanted the job, yet she still wanted to be part of Bell Labs. Happily, an arrangement was worked out that allowed her to remain as a part-time "distinguished member" of the technical staff at Bell Labs while teaching and doing research at Rutgers. There, she carried on her mission of nurturing future scientists by building a research group of young people and making them partners in her research.

> Shirley's life was about as good as it could be and there seemed to be nothing more that she could want.

But Shirley wasn't done yet. She also served on the board of a large utility company in New Jersey that owned or co-owned five nuclear power plants. For a time, she chaired the nuclear oversight committee of that board. In that job, she called on her experience in elementary particle physics, an outgrowth of nuclear physics.

During this same period, Shirley was also asked to be on the advisory council of an industry group that wanted to improve the operating performance of nuclear power plants. As she had at MIT, she impressed the members with her levelheadedness, her diplomacy, and her good judgment. Often, the members engaged in heated discussions and flung crazy arguments back and forth. And what did Shirley do? She sat quietly and listened.

Finally, when it seemed like nobody could agree on anything, she would interrupt the argument and ask, "What is it that you really want to do?" That made everyone stop in their tracks and examine the real issues. People would eventually come to an agreement.

Shirley's life was busy and full. But she enjoyed every minute of it. The work she did on the boards and commissions was challenging, and she got to meet interesting people and engage in stimulating conversations. But she still found time to nurture her family life. She was often able to squeeze in tennis matches with her husband and ski trips with her son.

Shirley's life was about as good as it could be and there seemed to be nothing more that she could want. Then, in 1994, the White House came calling.

By Shirley's second month
as **chairman** of the NRC,

she realized
she had walked into a **mess**.

A PRESIDENTIAL APPOINTMENT

9

S hirley was in her office at Rutgers University when she got a phone call from the White House. On the line was the director of the Office of Presidential Personnel. She had a question for Shirley: Would she like to submit her résumé for a presidential appointment?

"What kind of an appointment?" Shirley wanted to know. But the personnel director wouldn't tell her.

"Well, I can't send my résumé, then," Shirley said and ended the conversation.

A day or two later, the personnel director called Shirley again. "You're being considered for a job," she told Shirley, but she still wouldn't say what kind of job. This time Shirley agreed to send her résumé. She had to admit that she was intrigued.

Shortly thereafter, Shirley was summoned to Washington, D.C., to interview for a top-level job with the Nuclear Regulatory Commission (NRC). The NRC is an independent federal agency responsible for ensuring the safety of the nation's 110 nuclear power reactors. In addition, it regulates the uses of nuclear materials in medical, industrial, and academic settings and in facilities that produce nuclear fuel. It also oversees the transportation, storage, and disposal of nuclear materials and waste.

Shirley is sworn in as Chairman of the Nuclear Regulatory Commission *(above)*. Afterwards, she poses for a picture in her new office *(opposite)*.

Shirley was originally interviewed for a position as one of five commissioners. After her interview, however, the job offer got even better. President Bill Clinton instead offered her the job of chairman of the NRC. The chairman is the principal executive officer and the official spokesperson for the NRC. Shirley's public policy work in New Jersey and her work on the board of a utility company that owned nuclear plants gave her invaluable experience. And given the fact that Shirley had a Ph.D. in high energy physics (which is closely related to nuclear physics), it soon became obvious that she was uniquely qualified for this assignment.

But there were still a couple of hurdles to jump over. First, she had to submit to an FBI background check, which she passed easily. Then she faced a more daunting task. She had to be confirmed by the United States Senate, which meant she had to appear before a committee to answer questions. But being in the spotlight was not new to Shirley. Her poise, knowledge of the subject matter, experience, and strong resolve impressed the committee. Shirley was unanimously approved.

In 1995, under President Clinton, Shirley Ann Jackson became the first woman and the first African American to be appointed as

At Shirley's swearing-in ceremony, she gave a speech about her goals as Chairman of the NRC *(below)*. Shirley smiles for a "photo op" with Vice President Al Gore, her husband Morris, and son Alan *(below right)*.

chairman of the NRC. She had a staff of 3,000 people, a budget of $500 million, and a job that required many important decisions. But her first decision was a more personal one: Where would she live?

~ Commuter Mom

The NRC was in Washington, D.C., but Shirley had a home and family in New Jersey. Morris was still with Bell Labs, and Alan, at age 13, was excelling at his school. Shirley and Morris were both aware that this was a political appointment and she could very well be out of a job when the presidential administration changed hands. Together they decided that Shirley would stay in Washington during the week and commute home to New Jersey for the weekends.

During the week, Morris took Alan to school every day and then went off to work. Morris also picked Alan up after school, got him started on his homework, made dinner, and checked the homework. Shirley spoke with them both every night by phone.

During her visit to the Millstone Nuclear Power Station in February 1996, Shirley met with Senator Joseph Lieberman of Connecticut.

When Friday rolled around, Shirley arrived in New Jersey at 10 P.M. ready to devote the whole weekend to her family. Alan was on the swim team at school and also played water polo, so Shirley and Morris spent many weekends attending swim meets and water polo matches in New Jersey and all over the northeast.

Come Monday, Shirley was on the road by 5 A.M. to head back to Washington. As difficult as it was for Morris to be a "single parent" during the week, it was just as difficult for Shirley to be away from the day-to-day activities of her family. But they all felt that she was doing something important: protecting public health and safety when and where nuclear materials were used.

~ Cleaning Up the Mess

By Shirley's second month as chairman of the NRC, she realized she had walked into a mess. The public had raised many concerns about the safety of nuclear power plants, and the NRC had not done a good enough job dealing with those concerns. Shirley was a strong believer in nuclear energy as part of a diversified energy strategy for the United States. *(See box, page 86.)* But she also recognized that its success would be determined by how well the nuclear industry, the public, and the NRC could work together to ensure the safe use of nuclear energy.

The more she discovered about the situation, the more Shirley realized she had to do something to set things right. Never one to back down from a challenge, Shirley did what she had done back at MIT when she formed the Black Students Union: She gathered people together to discuss the issues. And again she insisted that everyone gather data to back up their claims.

Then, in March 1996, *Time* magazine ran a cover story called "Blowing the Whistle on Nuclear Safety." The article uncovered safety concerns about one of the units at the Millstone Nuclear Power Station in Waterford, Connecticut. The plant was sidestepping safety regulations, especially when it came to offloading nuclear fuel from the fuel rods. Nuclear reactors

During Shirley's tenure as Chairman of the NRC, she often visited nuclear power plants to discuss issues with workers and officials.

generate power in part through fuel rods, made from uranium. Periodically, these rods must be replaced with new ones. Because the rods produce dangerous radiation and are at a temperature of 250° F, they have to be disposed of very carefully. If proper procedures aren't followed, it can be disasterous. To make matters worse, the

Shirley observes workers handling fuel rods at a fuel fabrication facility, one of the NRC's areas of responsibility.

plant was not maintaining plant equipment in proper condition. And it had one more troubling problem: Workers who raised safety concerns were harassed or fired.

As soon as Shirley learned of the situation, she took swift action to correct the problems. First, she ordered the utility to enact an Employee Concern Program, which allowed nuclear workers to raise safety concerns without being harassed or threatened with job loss. She also launched a number of policies to improve training, accountability, and oversight among plant inspectors as well as among the NRC staff. Finally, Shirley ordered a review of all 110 nuclear reactors in the country to find out how many were moving fuel in violation of NRC safety rules.

Recognizing the importance of community, Shirley also reached out to ordinary people to hear their concerns. She held a public meeting in Connecticut to talk to residents and answer their questions. She made sure that the media was invited. Whenever she visited a nuclear power plant, especially a troubled one, she held a media briefing. She felt it was important for the public to be informed about what was going on regarding nuclear power plants.

During her four years as chairman, Shirley had to make many hard decisions. She toughened safety standards and fired or transferred senior managers who did not pay enough attention to safety issues. She brought in a more diverse management team.

Her evenhanded treatment of sensitive safety issues earned her respect from public safety watchdogs and members of the nuclear industry alike. Many people credit her with wisely resolving some of the toughest dilemmas the NRC has ever faced.

~ Going Global

Since Shirley had traveled all over the world, she had a larger world view than many folks. And she knew that nuclear safety was not something only Americans needed to be concerned about. It was a global issue as well.

Shortly after starting her job at the NRC, she had taken a trip to Chernobyl, Ukraine. The visit helped her realize the need for nuclear regulators worldwide to come together and discuss safety issues. In 1986, Chernobyl had suffered the worst nuclear disaster in history. A reactor steam explosion spewed highly radioactive

A Matter of Energy

Atoms contain huge amounts of energy known as nuclear energy. The strong forces that exist between particles in the nucleus of an atom create this energy, which can be released in two ways:

1. **Nuclear fission.** In this process a high-speed neutron is sent blasting through the nucleus of a heavy atom, which can then split into two parts, producing two or three more neutrons. This step is repeated, and a chain reaction takes place. Some of that energy is captured and used to generate power.

2. **Nuclear fusion.** In this process the nuclei of two or more light atoms stick (or fuse) together, releasing energy in the process. All the stars, including the sun, get their energy from nuclear fusion.

Whether energy is created through nuclear fission or nuclear fusion, some mass vanishes during the process. But where does it go? Albert Einstein showed through his famous equation $E=mc^2$ that this disappearing mass is converted into energy.

materials into the atmosphere. Much of the environment around Chernobyl was devastated, and many people became ill due to radiation exposure.

So in 1997 Shirley reached out to the international community and formed the first International Nuclear Regulators Association. This group included senior nuclear regulatory officials from Canada, France, Germany, Japan, Spain, Sweden, the United Kingdom, and the United States.

Shirley not only brought all these countries together, she also convinced them to get some real work done. The association created and accepted Terms of Reference, a document that lays out an organization's governing principles. The group also examined issues and offered assistance to other nations on matters of nuclear safety. Always the leader, Shirley was elected and served as the association's chair for two years.

JACKSON

As Chairman of the NRC, Shirley traveled all over the world representing the United States.

~ Physics and Public Policy

Shirley's background as a physicist greatly influenced her work as chairman of the NRC. Because she understood the science involved, she could make decisions based on well-established scientific principles. Also, being a physicist gave her credibility with those who worked in the nuclear industry, especially in other countries as she traveled abroad representing the United States on nuclear issues. Shirley was often able to look at a complex situation, find the key issues, and figure out a solution. Being a physicist meant that she was a problem solver and an organizer of material. She also had the ability to deal with complex situations.

But other aspects of her background helped, too. Her experience in organization and management at her sorority, at Bell Labs, and

The members of the International Nuclear Regulators Association, which Shirley *(fourth from right)* founded, pose for an historic photo in 1997. Shirley often signed bilateral agreements on behalf of the United States *(below)*.

KINDELAN   HARBISON   BISHOP   LAUPRAN   JACKSON   HENNENHOFER   TOGO

on corporate boards helped her in managing the NRC and developing good public policy in nuclear safety.

~ Another Phone Call

At this point in her life, Shirley felt she was making a difference in the world. Still, her term was coming to an end. She wondered what she would do when she had to move on. But President Clinton had other ideas. In the fall of 1998, he asked her to serve another term and continue to chair the NRC.

At about the same time, Shirley got another unexpected call. The chair of a search committee asked if she would be interested in becoming the eighteenth president of Rensselaer Polytechnic Institute (RPI). This school, located in Troy, New York, is one of the nation's oldest science and engineering research universities.

After considering hundreds of candidates, a 34-member panel unanimously chose Shirley to be president. RPI had enjoyed a long tradition of excellence, but had recently been plagued by a number of problems. The search committee members saw how successful Shirley had been in turning the NRC around. Now, they hoped, she could use her talents at the 175-year-old university and help make the letters RPI as recognizable around the globe as the letters MIT.

~ She Keeps Going…and Going…

As much as Shirley enjoyed her job at the NRC, she felt as though she had done about as much as she could there. Plus, she had to admit that she missed academic life. To her, shaping young people's minds and preparing them for the future was one of the most important things a person could do in life. Still, she had a number of things to consider before accepting the job.

As she had done with every other decision that would affect her family, she talked it over with her husband. Morris had always been supportive of Shirley's career and recognized that this was a tremendous opportunity. But it would mean moving from New Jersey to upstate New York. That meant Morris would have to give up his job at Bell Labs.

Shirley also needed to consider how this move would affect her son. Alan was in his senior year of high school, and Shirley did not want to disrupt his life any more than she had to.

Finally, she had one more consideration: President Clinton. He had asked Shirley to serve another term at the NRC, and she had not yet completed her first term, which would be over in the fall of 1999. The President's confidence meant a lot to Shirley.

Always the successful negotiator, Shirley managed to figure out a way to make it work. Although she was offered the RPI job in December 1998, Shirley got the Rensselaer board to let her start in July 1999. This allowed her son to finish his high school year. It also allowed Shirley to finish her term with the NRC.

At a farewell party at the NRC, her colleagues presented her with a special present. They gave her an Energizer Bunny™—a symbol that perfectly captured Shirley's boundless energy.

Now is your time.

Step through your window in time.

Look **forward**, not back.

Look up, not down.

Have **confidence** in yourselves.

10

BLAZING A TRAIL

A s Shirley Ann Jackson walked onto the stage of the Darrin Communications Center at Rensselaer Polytechnic Institute on December 11, 1998, a crowd of more than 800 rose to its feet in a standing ovation. It was still months before she would begin her position as president of RPI, yet already she had won over the students, faculty, and staff who had come to hear her speak. By the end of her speech, in which she laid out her plans for RPI's future, she had received a total of three standing ovations.

Two core tenets of her Rensselaer plan are diversity and community, which Shirley likes to call "communiversity." The term refers to the university as a community, as a family. It also emphasizes the fact that the university is a part of the city and of society in general.

Shirley's presidential inauguration on September 24, 1999, reflected this theme. The day's festivities included activities and entertainment on the Hudson riverfront in downtown Troy. To further emphasize the links between campus and community, Shirley kicked off the school year with a parade of students and faculty through the streets of Troy. The parade is still an annual event at Rensselaer.

Shirley views RPI as a family. She especially enjoys talking to the students (opposite) and sharing her ideas for the future. Above, Shirley takes a stroll around the RPI campus, shortly after her inauguration.

Shirley listens intently to a speaker at her inauguration as president of RPI. Shirley and Morris *(below)* pose in front of the president's house on the RPI campus. Morris retired from his job at Bell Labs and is now the Associate Director of the Center for Integrated Electronics at RPI.

As president of RPI, one of Shirley's first priorities was to increase the diversity of the student body. However, Shirley was smart enough to know that this could not be accomplished overnight. She knew that students would have to be reached early—as early as middle school—if they were to learn the fundamentals they would need to pursue a career in the sciences.

Shirley knew from firsthand experience that preparation to enter a scientific or technological career is a cumulative process. As she often tells people, "You cannot even begin to aspire to be a scientist, engineer, entrepreneur, or technological leader until you are comfortable with calculus. And you can't begin to think about calculus unless you've mastered algebra . . . and before that multiplication . . . and before that addition and subtraction."

Toward that end, Shirley has been a strong advocate of linking middle and high schools with colleges and universities. Under her leadership, RPI secured federal funding for a project that would help more than 900 local low-income students receive supplemental instruction in math, science, and technology, beginning in seventh grade and continuing through high school.

~ Expanding the Universe

Another current objective of Shirley's is to expand students' ideas about what they can do with a degree in physics. Typically, most people believe that having a Ph.D. in physics prepares a person to do one of two things: research or teach. But Shirley wants young people to understand that an education in physics can prepare them to do a number of jobs successfully.

Shirley's own varied career is a case in point. Whether she was examining the world at a submicroscopic level, studying the properties of matter, developing regulatory policy, or running a large university, she used her training as a physicist. Her focus has always been on making sound decisions by asking perceptive questions, making keen observations, and drawing insightful conclusions. A physicist is trained to do all of these things.

The three sisters—Gloria, Barbara, and Shirley *(left to right)*—take a short rest on a Yosemite hike.

~ Trailblazer

Since her days at CERN, Shirley has been an avid hiker. Once a year, she and her sisters go on a hike with a group of primarily African American women. All three sisters have excelled in their chosen professions. Gloria Joseph, the youngest, is a lawyer and director of administration at the National Labor Relations Board. Barbara Avery is vice president of student life at Holy Names University in Oakland, California. (Sadly, their brother, George, died in 1984.)

The idea behind the hikes is to get women to stretch themselves physically as well as mentally. On these hikes, the women see that they can do things they wouldn't necessarily do otherwise.

Barbara usually organizes the event. The group—which ranges from 3 to 16 women—has hiked in the Sierra Nevadas, in Yosemite National Park, at Mount St. Helens, in the Olympic Peninsula, in the Rockies, and at Lake Tahoe. Early on, Shirley earned the nickname "The General" for her ability to read a map

Shirley is still very close with her family. Here, Barbara, Shirley, Mom, and Gloria pose for a snapshot at a recent "Communiversity" event at RPI.

and for her fearlessness in leading the women down the right paths.

It hasn't always been easy. One time the group went hiking in the Sierra Nevadas. They came to a point where they had to cross a deep creek. As part of the trail, a large log had been thrown across the creek. Water rushed under the log, cascading down the mountain, making it dangerous if anyone fell. But there was no other way to get to the other side of the creek.

Shirley assessed the situation. She knew it was dangerous, but someone had to cross first. She gingerly took a step onto the log. At first she struggled for balance and nearly fell off. But by concentrating and moving slowly, Shirley soon regained her balance and walked across to the other side. The other women were impressed by how easy Shirley had made it look. They took her lead and followed.

Later that evening, Shirley admitted to them that she had been frightened. But she knew that if she hadn't set the example, the others wouldn't have gone across. Once again, Shirley's accumulation of life lessons shone through. She proved to herself and others how important it is to confront fears and work to overcome them.

That attitude is what has made Shirley a trailblazer in so many areas, from being the first in her family to graduate college to becoming the first African American woman president of RPI. But Shirley believes that being a trailblazer is only a good thing if one paves the way for others to follow. Shirley's life has been a shining example of nurturing and inspiring others to follow in her footsteps. And just as she overcame the struggles she faced in her early years at MIT by staying focused on her goals, she encourages young women—especially young minority women—to focus on what they want to do in life and not let their struggles hold them back.

This advocacy, coupled with a life of great accomplishment, was recognized by the National Women's Hall of Fame. In 1998, Shirley was inducted into this prestigious group for her work in

theoretical physics, nuclear safety, and education. There she joins such towering historical figures as Susan B. Anthony and Sojourner Truth.

Her attitude and passion make Shirley a speaker in great demand. She is often called upon to give commencement addresses at colleges and universities throughout the country. In May 2003, Shirley returned to her childhood home of Washington, D.C., to give a commencement speech to the graduates of the University of the District of Columbia. Looking out over the sea of hopeful faces, Shirley delivered an inspiring and passionate speech that spoke of her own challenges and opportunities. She encouraged the graduates to pursue excellence, to lead, to persevere, and to embrace people of all kinds. She said to them:

> Now is your time. Step through your window in time. Look forward, not back. Look up, not down. Have confidence in yourselves. Take care of yourselves and your families. And, when you are feeling tired, discouraged, or just plain disgusted, think of the bridges you already have crossed, the mountains you already have climbed. Do not let others set your aspirations for you. Set them yourselves, and work to achieve them. Intend to make a difference in this world in large ways and small.

Inspiring words from a woman who really *has* made a difference in this world and continues to do so.

Timeline of Shirley Ann Jackson's Life

1946   Shirley Ann Jackson is born on August 5 in Washington, D.C.

1954   After the Supreme Court of the United States knocks down segregation laws and orders schools to integrate their student populations, Shirley switches to the Barnard School, a neighborhood elementary school.

1964   Shirley graduates as valedictorian from Theodore Roosevelt High School in Washington, D.C.

1968   Shirley earns a bachelor of science degree in physics from the Massachusetts Institute of Technology (MIT). Shortly after beginning her graduate studies in physics at MIT, Shirley forms the Black Students Union to help change the school's policies and practices toward African American students.

1973   Shirley becomes the first African American woman to receive a Ph.D. in any field from MIT. She begins a postdoctoral position at the Fermi National Accelerator Laboratory in Batavia, Illinois.

1974   On a Ford Foundation grant, Shirley becomes a visiting science associate at the European Organization for Nuclear Research (CERN) in Geneva, Switzerland.

1976   Shirley completes her postdoctoral work at Fermilab. She becomes a member of the technical staff at AT&T Bell Laboratories in New Jersey.

1979   Shirley weds fellow physicist Morris A. Washington.

1981   Shirley and Morris's son, Alan, is born.

1985   New Jersey Governor Thomas Kean appoints Shirley to the state's Commission on Science and Technology.

1986   Shirley is elected a Fellow of the American Physical Society.

1990   New Jersey Governor James Florio awards Shirley the Thomas Alva Edison Science Award for her contributions to physics and the promotion of science.

1991   Shirley begins teaching as a professor of physics at Rutgers University. Meanwhile, she remains a "distinguished member" of the technical staff at Bell Labs. Shirley is elected a Fellow of the American Academy of Arts and Sciences.

1993   Shirley is awarded the New Jersey Governor's Award in Science.

1995   President Bill Clinton appoints Shirley as Chairman of the Nuclear Regulatory Commission. She becomes the first female and the first African American to head the agency.

1997   Shirley forms the International Nuclear Regulators Association and is elected as the group's first chair.

1998   Shirley is inducted into the National Women's Hall of Fame.

1999   Shirley becomes the 18th president of Rensselaer Polytechnic Institute (RPI) in New York. She is the first African American woman to lead a national research university.

2000   The Women in Technology International Foundation inducts Shirley into its Hall of Fame.

2001   Shirley receives numerous awards, including the Richtmyer Memorial Lecture Award from the American Association of Physics Teachers, the 15th Annual Black History Makers Award, and the Black Engineer of the Year Award. She is the first African American woman to be elected to the National Academy of Engineering.

2002   *Discover* magazine names Shirley one of the Top 50 Women in Science.

2004   Shirley serves a one-year term as president of the American Association for the Advancement of Science.

2005   Shirley continues to serve as President of RPI.

About the Author

Diane O'Connell is the author of five books, including another biography in this series, *People Person*. Although physics is a new subject for her, Diane is used to tackling complex scientific issues, as seen in her award-winning medical reporting on the topics of multiple sclerosis, hemophilia, and gene therapy. Diane and her book *Divorced Dads: Shattering the Myths*, which she coauthored with Sanford L. Braver, Ph.D., were featured on the news program 20/20. Before writing books, she was on staff at Sesame Workshop as a writer and editor for *Sesame Street Magazine*. Diane lives in New York City with her husband Larry and golden retriever Palmer.

Glossary

This book is about a theoretical physicist, a scientist who doesn't do experiments but uses mathematics to model the world around her. To figure out the meaning of scientific words, it helps to know a little Greek and Latin. The word physics comes from the Greek *physis*, meaning "nature" and the Latin *physica*, meaning "natural science." Physics is a science that deals with the physical properties of matter and energy. The word theory comes from the Greek *theorein*, meaning "consider" and *thea* meaning a "sight or view." So a theoretical physicist comes up with a theory or possible solution to a problem, which is then tested in a laboratory experiment.

Here are some other scientific words you will come across in this book. For more information about them, consult your dictionary.

aerodyamics: a branch of physics that deals with how gases in motion and forces (such as air pressure) affect objects moving through air. Aerodynamics is an important consideration in the design of airplanes, automobiles, and even go-carts.

atom: the smallest particle of matter that still has its properties. Each atom is made of even smaller parts: protons and neutrons in a nucleus at the center of the atom and electrons that surround the nucleus.

conductor: any material that lets electric current pass through it. A semiconductor is material, such as silicon, that conducts electricity only under certain circumstances and is used to control the flow of electricity in electronic devices. A superconductor is material, such as lead or tin, which, at temperatures near absolute zero (the lowest temperature that is possible), allows electrical current to flow without resistance.

crystal: a solid composed of atoms that are arranged in orderly, repeating patterns

crystalline: made up of crystals or solidified in the form of crystals. In a multicrystalline form, small crystals are arranged haphazardly, although the atoms in each small crystal are highly ordered.

energy: the power to do work. Heat, light, and electricity are different forms of energy. The word comes from the Greek *en*, meaning "in" and *ergon*, meaning "work."

gravity: the natural force that causes attraction between any objects having mass. The word comes from the Latin *gravis*, meaning "heavy."

mass: the amount of matter an object contains

matter: anything that occupies space, has mass, and exists as a solid, liquid, or gas

neutrino: a subatomic particle that has no electrical charge and virtually no mass. It can pass through any solid material and emerge virtually unchanged. Because it is electrically neutral, a neutrino rarely interacts with other particles.

nuclear energy: a powerful form of energy produced within the nuclei of atoms. It is released by fission (splitting the nucleus of an atom) or fusion (combining the nuclei of two or more atoms).

particle: any of the tiny units that make up matter, such as an atom, a proton, or an electron. A subatomic particle is any particle found within an atom. The word comes from the Latin *partem*, meaning "part."

quantum mechanics: a branch of physics that studies the tiniest parts of matter. The word quantum means "how much" in Latin.

quark: a subatomic particle. There are six known types of quarks believed to make up protons and neutrons. The six quarks have whimsical names—they are called up, down, charm, strange, top, and bottom.

theory of relativity: a theory relating to the concepts of matter, space, time, and motion expressed in certain equations developed by Albert Einstein

transistor: a small electronic device, which contains a semiconductor that controls the flow of electrons in an electric circuit and can work like an electronic switch

tunneling: how electrons pass through materials when there are barriers to their movement

---

Metric Conversion Chart

When you know:	Multiply by:	To convert to:
Inches	2.54	Centimeters
Miles	1.61	Kilometers
Acres	0.40	Hectares
Centimeters	0.39	Inches
Kilometers	0.62	Miles
Hectares	2.47	Acres

FURTHER RESOURCES

Women's Adventures in Science on the Web

Now that you've met Shirley Ann Jackson and learned all about her work, are you wondering what it would be like to be a physicist? How about a planetary astronomer, a forensic anthropologist, or a robot designer? It's easy to find out. Just visit the *Women's Adventures in Science* Web site at **www.iWASwondering.org**. There you can live your own exciting science adventure. Play games, enjoy comics, and practice being a scientist. While you're having fun, you'll also get to meet amazing women scientists who are changing our world.

BOOKS

Bryson, Bill. *A Short History of Nearly Everything.* New York: Broadway Books, 2003. This engaging, often funny, always illuminating romp through space and time seeks to answer such elemental questions as: "How small is an atom?" "What is a black hole?" "How and when did the solar system get started?" A great read for all ages.

Hinds, Patricia M., ed. *Essence: 50 of the Most Inspiring African-Americans.* New York: Essence Books, 2002. Gorgeous photographs fill this book profiling people of color who are making a real difference in the world. In addition to Shirley Ann Jackson, the book features inspiring stories about Michael Jordan, Oprah Winfrey, Condoleeza Rice, and Venus and Serena Williams.

McKeever, Susan, et al, eds. *The DK Science Encyclopedia*, London: DK Publishing, 1998. No bookshelf should be without this excellent science resource. You'll learn everything you ever wanted to know about matter, energy, space, ecology, the Earth, how living things work, and so much more. Fascinating illustrations and photographs fill every page.

WEB SITES

Fermilab: www.fnal.gov
Take a virtual tour of the famed high energy physics laboratory, get answers to frequently asked questions about physics, and learn amazing things about the most basic particles and forces of nature.

The Particle Adventure: http://particleadventure.org
This site takes you on an adventure of the entire known subatomic world, from accelerators to quarks, using illustrations and understandable explanations.

CERN: http://public.web.cern.ch/Public/Welcome.html
Learn everything about the European Organization for Nuclear Research, one of the world's largest particle physics centers. Find out how and why physicists study particles and how particle research is useful in everyday life. You'll also find fantastic pictures of particle collisions.

SELECTED BIBLIOGRAPHY

In addition to interviews with Shirley Ann Jackson, her family, and colleagues, the author did extensive reading and research to write this book. Here are some of the sources she consulted.

Bryson, Bill. *A Short History of Nearly Everything.* New York: Broadway Books, 2003.

Cole, K.C. *The Universe and the Teacup: The Mathematics of Truth and Beauty.* New York: Harcourt Brace & Company, 1997.

Feynman, Richard P. *Six Easy Pieces: Essentials of Physics Explained by Its Most Brilliant Teacher.* Cambridge, Massachusetts: Perseus Publishing, 1998.

Frank, David V., Ph.D., et al. *Science Explorer: Physical Science.* Needham, Massachusetts: Prentice Hall, 2002.

Gamow, George and Russell Stannard. *The NEW World of Mr Tompkins* [sic]. Cambridge, United Kingdom: Cambridge University Press, 2001.

Gribbin, John. *In Search of Schrödinger's Cat: Quantum Physics and Reality.* New York: Bantam Books, 1984.

Hinds, Patricia M., ed. *Essence: 50 of the Most Inspiring African-Americans.* New York: Essence Books, 2002.

Kane, Gordon L. *The Particle Garden: Our Universe as Understood by Particle Physicists.* New York: Basic Books, 1995.

McKeever, Susan, et al., eds. *The DK Science Encyclopedia,* London: DK Publishing, 1998.

Veltman, Martinus. *Facts and Mysteries in Elementary Particle Physics.* River Edge, New Jersey: World Scientific, 2003.

INDEX

Illustration Credits:

Illustrations: Max-Karl Winkler

The border image used throughout the book is of particle tracks from 1973 and can be found at doc.cern.ch, image number 11465.

JHP Executive Editor: Stephen Mautner

Series Managing Editor: Terrell D. Smith

Designer: Francesca Moghari

Illustration research: Joan Mathys

Special contributors: Meredith DeSousa, Mary Kalamaras, April Luehmann, Kate Jerome Nyquist, Anita Schwartz, Leanne Sullivan

Graphic design assistance: Michael Dudzik and Anne Rogers